W9-BJV-343

The Technique of Casting for Sculpture

The Technique of Casting for Sculpture

John W. Mills

B. T. Batsford Ltd, London

First published 1967

ISBN 0 7134 6157 8

Typeset by Servis Filmsetting Ltd., Manchester
and printed in Great Britain by
The Bath Press, Bath
for the publishers
B. T. Batsford Ltd.
4 Fitzhardinge Street
London W1H 0AH

Contents

Acknowledgments

To acknowledge personally all the people who, and all the places that have helped me in my understanding of the skills involved in the production of a work of sculpture is an almost impossible task. I can only thank those persons, and be grateful to those museums, art schools, workshops and studios that have contributed in various ways to my education in these matters. Good teachers do not harbour petty secrets of technique and method, and I was fortunate always to have good teachers. I hope that with this book I can in turn make a contribution to the proper appreciation of skills and techniques in sculpture.

I am indebted to Lynton Money for his excellent photographs. He made 56 of the illustrations, most of them taken under very trying conditions. Their clarity marks his understanding and his skill. I am also grateful to Mr Jean Koefoed of Reinhold who encouraged the idea for this book during my visit to New York in 1965; and to Miss Thelma M Nye of Batsford who has nursed it patiently through the complexities of publication.

Hinxworth Place John Mills

Introduction

Techniques of casting for sculpture have evolved over centuries and are unlikely to change dramatically despite the emergence of materials peculiar to the twentieth century. I have written this book, however, to provide details of all the casting techniques employed by sculptors to produce work in various materials.

It is safe to say that carving is a straightforward method of producing an image. Sophisticated techniques of fine carving are, of course, diverse and extremely complex, and watching an experienced carver at work only emphasises this point. The fact remains, however, that by simply using a knife or chisel it is possible to carve a crude image. This basic fact presents itself to almost every small boy with a knife as immediately as it did to prehistoric man, who used such methods to make most of his tools and weapons, chipping or rubbing away material until he had shaped his axe, spear or arrowhead, needle or knife. In this sense the act of carving is basic and elemental. Casting is not.

Teaching students at various levels confirms my belief that modelling, and the necessary casting skills appertaining to modelling media, must be taught with care. With the exception of casting using high explosive (a prohibitively expensive business) no new moulding and casting method has been devised that does not require the understanding and practice of the basic principles of mouldmaking and casting. My proposition to all students is consistently this: practise, absorb and understand the basic skills to such a degree as to enable you to tackle the most complex casting problem with confidence.

Given clay, students will at first more often than not cut away unwanted material rather than use the process of addition to build an image. The resulting forms are conceived without finesse and are usually more suited to carving than modelling and its associated technique of casting, which appears as a mysterious and intimidating procedure. It is not an obvious process in that it does not immediately present itself as the method best suited to making sculpture of finesse allowing for the greatest freedom of form. It is a technique which has to be taught.

The simple geometric shapes which I suggest making to practise on are a good method for grasping moulding processes quickly. Some impatient people will endeavour to begin casting using a sculpture they have spent a great deal of time and effort modelling, and this will inevitably cause undue pressure to avoid mistakes – in such cases, the greater responsibility is placed on the shoulders of a tutor or advisor, and the learning will take longer. You cannot cast with ease having not made mistakes and thus gained experience. It takes time to gain the skill with which you can assess complex form and volume and relate these to the mechanical procedures of casting. It requires experience to make a mould that divides a complex volume adequately and allows the original, in whatever form, to be removed from the mould and replaced by the casting medium. I did have a graduate student at the University of Michigan, who successfully completed three life size moulds and high aluminous cement casting having seen no practical demonstrations. Circumstances were such that I could only describe to him, using diagrams and detailed instructions, the process involved, leaving him with a sheaf of notes and drawings after each tutorial with which to proceed to the next stage. The results were excellent, but I consider him to be a rarity and do not recommend his approach except in peculiar circumstances.

When I wrote my first textbook, *Sculpture in Ciment Fondu*, I did so because other sculptors and students were anxious to know about this particular technique developed by myself and a colleague, Sydney Harpley. This casting technique, using ciment fondu, made it possible for sculptors to produce in their studios a sculpture able to withstand the weather and, at the same time, strong enough to be practical for placing in public sites. Young sculptors could complete work – a commission, for instance – at a reasonable cost and see their work placed, thus gaining experience and the possibility of making the money necessary to make more sculpture.

It is important that a specialist, and the artist is most certainly a specialist, should know as much about his work as possible. Casting is an important technique to a sculptor, but it does not follow that, because he knows how to do the work himself, he will not employ a caster. What it does mean is that he will know what to look for when the casting is made, and the forms he designs will completely embrace the total craft of cast sculpture. To have little or no knowledge means the sculptor works only up to the point where the caster takes over; what goes on after that remains a mystery. This can lead to conflict, demands made on the caster may be impossible or the caster may be inadequate to resolve particularly complex images.

Constantin Brancusi was a sculptor with a working knowledge and good understanding of technical methods involved in producing a cast bronze sculpture, and he used this knowledge to make unique and beautiful statements in sculpture. Similarly, Henry Moore's later bronze sculptures have a unique beauty that came from an acute appreciation of the material and the techniques involved. Such knowledge gives a sense of freedom to develop a personal image, and such freedom is evident in all great works. It is perceivable in Egyptian and Greek sculpture, in Indian and Chinese, and in the works of Donatello, Michelangelo and Rodin. My favourite quotation is that splendid utterance of Rodin's, 'one must have a consummate sense of technique to hide what one knows'. Technique is purely the technical processes by which a form or image is achieved. Dürer, so the story goes, was once asked to show another painter the brushes he used to paint the fine delicate hair that is so typical of his portraiture. He agreed to do so, and duly arrived at the studio of the painter concerned; picking up one of the many brushes there, he proceeded to demonstrate, proving that no special brush was necessary, but rather a complete understanding of the materials to hand. This allows the mind and imagination absolute freedom to invent and develop according to its potential.

Understanding a technique does not, however, mean a stereotype 'do-it-yourself' method or formula by which one can model a form. Neo-classical painting and sculpture were often produced in such a restricted way in an attempt to reach the high standards set by the Greeks and Romans, but a lack of understanding of the mental freedom that these ancient sculptors possessed resulted in a barren period for art. Each original sculpture presents a

1 and **2** *Two-piece reclining figure No. 4* by Henry Moore, 1963, bronze. Original version of this figure was made for the Lincoln Centre, New York.

3 (Top) *Two-piece reclining figure No. 1* by Henry Moore, 1959, bronze. *Chelsea School of Art, London*

4 (Bottom) *Two-piece reclining figure No. 5* by Henry Moore, bronze.

unique set of problems during the manufacturing process and each of these problems must be assessed and dealt with on a one-off basis. Repetition casting for editions, where each casting needs to be identical, is the exception that proves the rule. The practiced sculptor must use his skill and experience as a guide to solving each problem individually, whilst all the time considering the various techniques and procedures that can effect the piece.

I believe there should be no secrets regarding technical processes in art. The means of making manifest a particular personal image should, as far as possible, be made available to all those who require such knowledge. Casting is, of course, only part of the production of a sculptured image and is subordinate to the creative process, but it remains a vital skill that can encourage others to give a greater freedom of expression.

5 and **6** *Three-way piece (Points)* by Henry Moore, 1964, bronze. This sculpture was photographed at the Greater London Council's 1966 *Sculpture in the Open Air* Exhibition at Battersea Park, London.

I Casting

What is casting? This is a question sculptors are often asked of what is, in fact, one of the most widely used techniques in sculpture. Every sculptor has at some time to use a casting method, be it crude and quick or sophisticated and involved. Be he a carver, modeller or assembler, at some point he finds it necessary to make a mould, then from that mould to produce a casting. This process has been used almost since the beginning of time.

The mould may be fashioned in a primitive way, such as the way primitive man carved in stone a cavity the shape of an arrow or spearhead, into which cavity he poured molten metal to make in fact a casting. Every kitchen today is equipped with just such a primitive mould – the domestic jelly mould. This embodies the basic mould-casting principle.

The mould may be an extremely complex cavity such as used in the plastics industry, to produce a wide range of complicated moulded products. In reverse it is the same pattern and shape as the product, incorporating a certain amount of plumbing to cool the heated plastic that is forced under great pressure into the mould cavity. The cooling hardens the plastic instantly which can then be removed from the mould, revealing its form, such as a milk crate or a plastic doll, in an endless production run.

In their extremes these two methods demon-

7 and **8** Sumerian terracotta press mould and the casting made from it. *British Museum*

strate the opposing poles of achievement in the history of casting. They demonstrate also the fact that the basic procedure is the same for all castable substances. The procedure differs only in detail according to the properties of the material and the ultimate use of the casting.

The earliest and most primitive kind of mould, which it is possible to trace, is the open mould. This was made by carving the negative shape in a soft stone, or by modelling such a shape in clay to be baked to make a terracotta mould. From such a mould, castings in various materials could be made. The earliest of these would be mud or clay. Later wax castings were made and, of course, metal. In this way, cast artefacts could be produced in number. Any substance that could be poured in a liquid state and would harden, could be used in such a mould, and this is the common denominator of all castable materials. The British Museum has a collection of Sumerian terracotta moulds from which clay press castings were made. These illustrate the basic technique that is common to all cultures, similar evidence of which exists in museums all over the world.

Another thing almost all cultures have in common is the skill of metal casting. All civilizations seem to have developed the art of lost wax bronze casting, plus the diverse uses of terracotta, both as a moulding material and as a casting material. Gypsum, a form of plaster, is common to all the more sophisticated cultures. We know for instance the Greeks and Egyptians used it. Egyptian sculptors made plaster life and death masks from which they worked on their monumental portraits; the masks being always available, whereas the subject would not be. This was a practice also used by the Roman sculptors. The plaster was a fairly coarse material, made by grinding gypsum or chalk to a powder that was mixed with ground marble dust, and glue size was added as the binding agent. It was not a material for refined work, simply an expedient means of making a model from which to work on the carving.

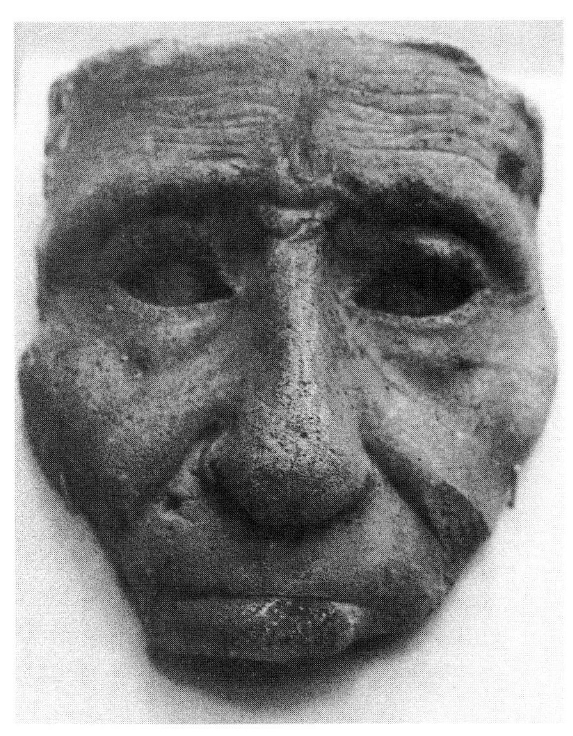

9 Egyptian plaster cast taken from life. Such masks provided a constant reference for the sculptor. *British Museum*

Of the sculptors' materials, other than stone or wood, terracotta and metal alloys have been the most important, and form a nucleus around which the sculpture of most cultures has developed. Today we can add to these traditional materials the innovations and refinements of media arising from industrial demands and experiments. Gypsum has been superseded by plaster of Paris and its further refinements. Cements and concretes have been made to suit the sophisticated requirements of modern building and are, of course, necessarily more refined and stronger. Resins and many various synthetic casting materials are also included. The basic principles have not altered; we still use the same methods of casting as the Greeks and Egyptians, Romans and Chinese, but our skills are more refined. We have castings of much greater accuracy than ever enjoyed by sculptors in the past. The earliest known bronze cast is Indian, dated at about 5000 BC, and we must assume a previous development period of the lost wax bronze casting method used then. This method is the same as that used still, but with the difference of superior moulding techniques, made possible by the production of refined refractory plasters for investments, and flexible moulding materials. These developments make it possible to retain the original sculpture (master cast) so that there is no risk of the total loss of the work should the metal fail to fill the mould properly. Formerly the original was made directly in the wax, which would then be covered with a mould, the wax melted out leaving a cavity that would subsequently be filled with molten metal. This meant that the sculptor had only one shot at the pouring, and if it did not pour properly it was lost completely, which could mean the total loss of months of work.

Metals were for centuries the only durable casting substance; terracotta being the next best, but always vulnerable. Sculptors were, therefore, expected to have a working knowledge of metals in order to be able to chase the metal to a fine surface. With the Industrial Revolution came improvements in materials and methods, enabling quicker and more uniform work. If you examine a typical Indian bronze sculpture, you will see the tremendous amount of work put into the surface and, by chasing, forms have been fully realised; jewellery engraved, and drapery decorated. The form and detail of the sculpture are typical of

10 *The Thinker*, from *The Gates of Hell* by Rodin, bronze. *Rodin Musem, Philadelphia Pennsylvania*

11 Detail of *The Thinker*

those modelled with wax, to be refined after casting.

Compare such a sculpture with Rodin's *The Thinker* (*10* and *11*) and look at the quality of the surface according to the form, and the form according to the surface. Rodin was free from the necessity of chasing the bronze cast. He modelled freely with clay concentrating on the form, the sensuality of surface, the personal imagery, knowing he could be confident in the caster's ability to reproduce in bronze exactly what he had modelled. The forms and marks he made would be cast faithfully, first in plaster of Paris (to produce the master cast from which the flexible mould would be made), to make the wax positives, which in turn would be made in bronze. Rodin's sculptures would be sent from his studio bearing all the marks of the master, including his fingerprints, and would return as an accurate reproduction in bronze. Perhaps Michelangelo would have been more favourably disposed towards bronze if he could have enjoyed the certainty of casting processes as did Rodin. The forms made by Rodin and by most sculptors since have been determined mostly for their own sake without compromise, or allowances being made for inhibiting technical considerations. There are some people who deplore this situation, regarding the dry controlled technique of oriental sculpture as more desirable than the lack of inhibiting restrictions.

The point is that sculptors today have the choice. This surely is what is so exciting about development and innovation. Technicologically and aesthetically sculptors have an almost absolute freedom, and it is up to the sculptors of the twentieth century to justify this position by the quality of their sculpture.

To study the techniques and methods of sculpture, the student should stop simply admiring images and try to see how these images have been made. Try to identify the seams, the runners, the chase marks and the marks left by the various tools used by the

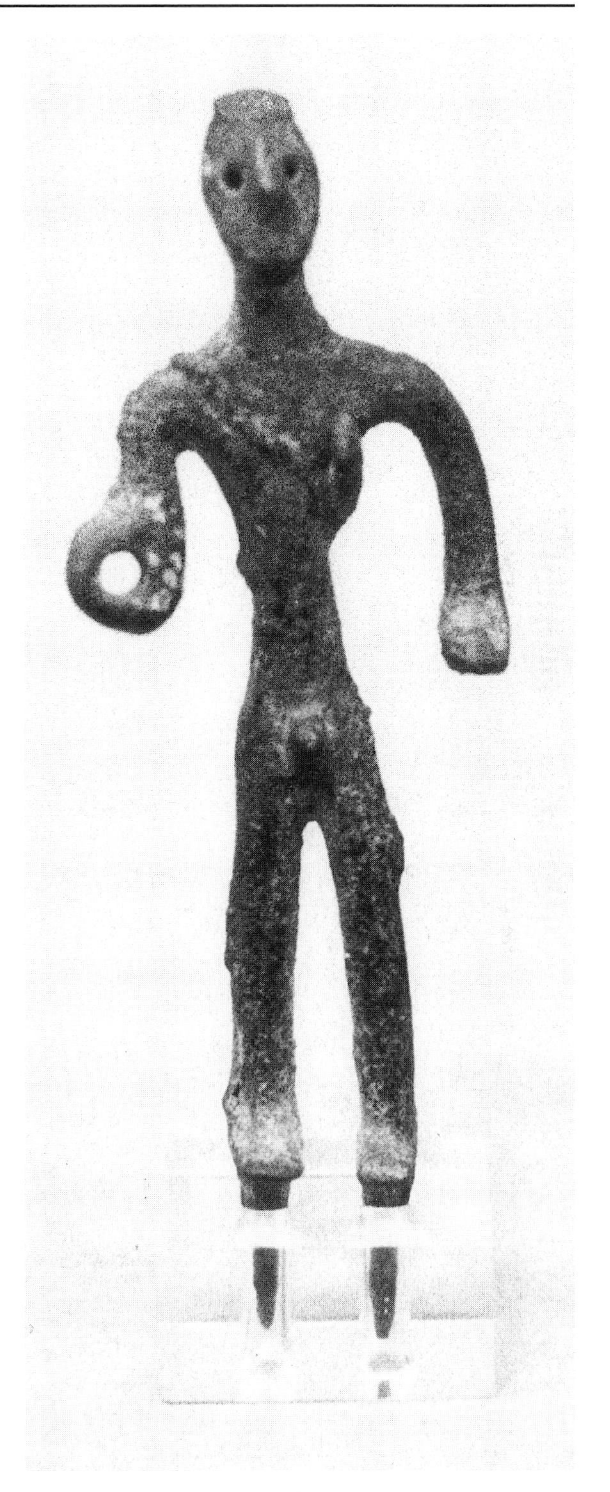

12 Warrior 700 BC, found near Peloponnes. *British Museum*

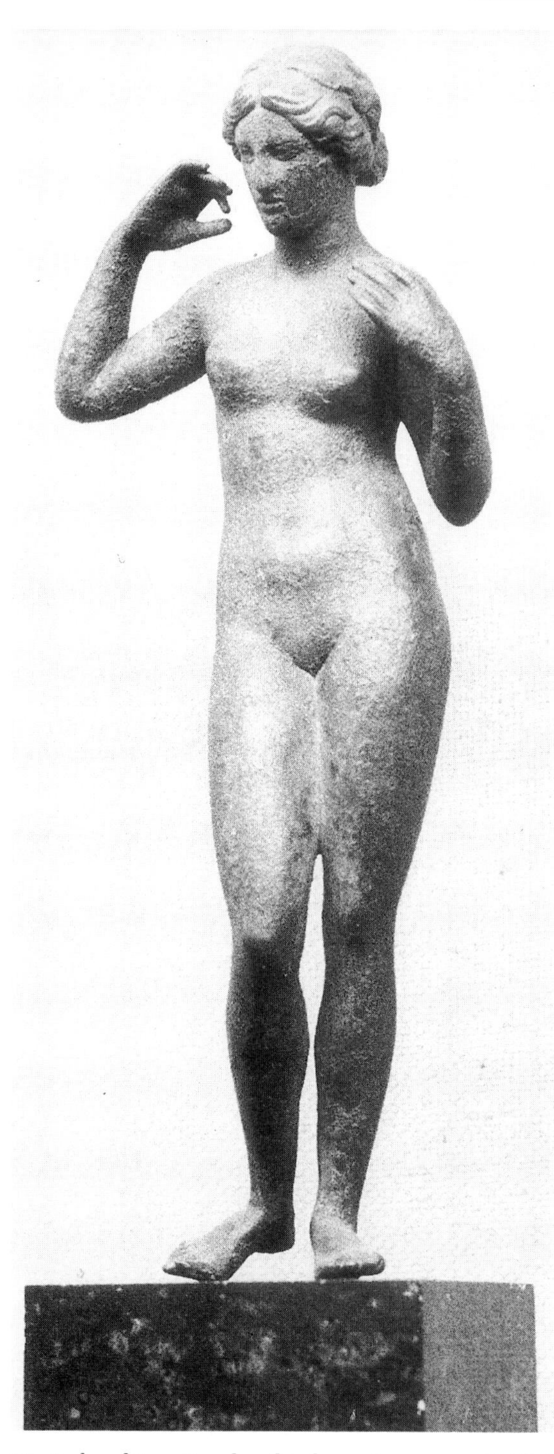

sculptor. Look, as it were, under the skin of sculpture, determine what methods and materials have been used, discover what technical thought went into the manufacture of the work. It may often be impossible, as indeed many sculptors are concerned to cover up all traces which might mean the visual destruction of a sculptured form. Seam lines, runner or pin marks often occur undesirably and do not enhance the form but obscure it at that point.

If you examine an early Greek bronze (12), you will see that this has been made directly in the wax, then cast in bronze. The shapes have been achieved by rolling the wax to form arms and legs and the torso, decoration has been made in the same way, simply and directly. The sculpture we see is almost exactly as it came from the investment mould. These kinds of bronzes can be regarded as examples of the most basic bronze casting technique produced by all societies.

Next examine more sophisticated Greek bronzes (*13*), some of which were made by exactly the same method as the earlier sculp-

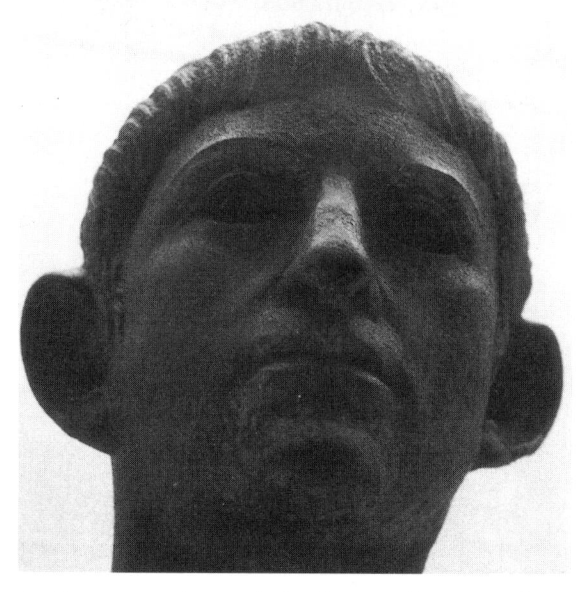

13 Aphrodite, Greek, third century BC. *British Museum*. This is much more sophisticated than the warrior in figure 12, which was modelled direct from wax and then cast.

14 Bronze portrait of the Roman Emperor Claudius, probably cast by the lost wax technique but with a great deal of work carried out on the final bronze. *British Museum*

tures; a similar technique indeed was used by the sculptors of the Indus Valley civilization. Forms larger in volume, figures of large proportions, demanded more sophisticated methods of casting. These forms had to be made hollow; the basic reasons for this are firstly the obvious one of weight, secondly the shrinkage of the metal upon cooling, which demands a fairly even thickness of metal throughout. Uneven thickness causes severe cracking, due to uneven shrinkage. These sophisticated and more complicated methods took a great deal of time to complete, and the sculptors were concerned to evolve a form commensurate with the great effort involved, and consequently fine chasing and intricate painstaking work was done on the casting.

Study the early bronze casts and try to trace the development from the simple solid cast to

16 and **17** Bronze weights from Ashanti, 2.5cm (1in) high, modelled direct in wax and cast by the lost wax process. *In the collection of Nan Youngman*

15 Bronze mask from Ashanti. *In the collection of Nan Youngman*

18 Bronze weight from Ashanti. A locust reproduced in bronze by the lost wax process. The actual insect was invested and burned out of the mould to provide the cavity for the metal to fill. *In the collection of Nan Youngman*

the hollow casts. All civilizations practised in the art of sculpture, offer the diligent student cryptic illustration of the development of techniques used in foundries right up to the Industrial Revolution. By looking carefully at these sculptures a great deal can be learned, particularly with regard to how far a clay, plaster or metal surface can be taken prior to bronze casting. Brancusi, for instance, made some startling forms in polished metal, casting from a smooth marble original, to exploit in polished metal its nuances of reflection and light. Benin or Ashanti bronzes on the other hand were realised in the wax state, the bronze being simply an accurate reproduction of that image (*15* to *18*).

When examining larger metal sculptures, it will be noticed that some of the forms appear coarse. Shapes are not so crisp, the surface not quite so close. These usually are castings taken from sand moulds, a technique known to have been widely used by the ancient Greeks. The sculptures are less delicate, made with little or no undercut form that might drop from the sand when the hollow mould is awaiting the bronze pouring. Today innovations of hardening binders and additives enable sand moulds to be much more intricate, using, virtually, the method of piece moulding with sand. This involves the manufacture of separate pieces of mould for every undercut.

Metals, because of their permanence compared with most other materials, have preoccupied sculptors almost to the exclusion of other materials. This is, I think, still the case, and metals are always treated with greater respect than other casting media.

Historically, clay is the next most important castable material, but it is used much more extensively as a simple modelling material. Cast terracotta, in the main, has been limited to the field of production from a master mould of components which can be assembled, to make items such as dolls, repetitive figures and domestic ceramic vessels. Primitive and ancient civilizations used the technique of casting from a master mould for the manufacture of easily recognisable images of their deities. Usually the cast parts were the heads and limb extremities, the area between being modelled directly and the image-character established by the addition of the cast components. Mexican terracotta figures and dolls, examples of which most museums include in their collections, were usually made using such methods. The British Museum possesses case upon case of fine examples of such Mexican sculptures, their techniques clearly visible to the searching eye. Casting from a master mould was common to almost every culture, examples of which are to be found from the cultures of Greece, Egypt and China. The Royal Ontario Museum at Toronto has a very fine collection of Chinese funerary sculptures, many of which demonstrate this method, plus every other technique of production possible with fired clay. There can be seen numerous showcases displaying complete arrays of figures symbolising whole households, from the most menial servant to the highest ranking officers.

In the sphere of ceramics, vessels are often cast from a mould. Such complicated items as teapots are slip cast. Industrially these techniques are refined to an almost ridiculous degree, making possible the mass production of hundreds of items per day. Industrial techniques on this scale have almost eliminated the aesthetic concern for shape and form and are to blame for the often poor quality of shape prevailing in the ceramics industry. The mould and ease of production run is given

19 *Mask of Van Rijn* by John W. Mills, bronze. Produced by the lost wax process from an original modelled direct in wax.

20 *Head of Van Rijn* by John W. Mills, sand cast aluminium. This head was made by producing three pattern sections that were cast and then bolted together.

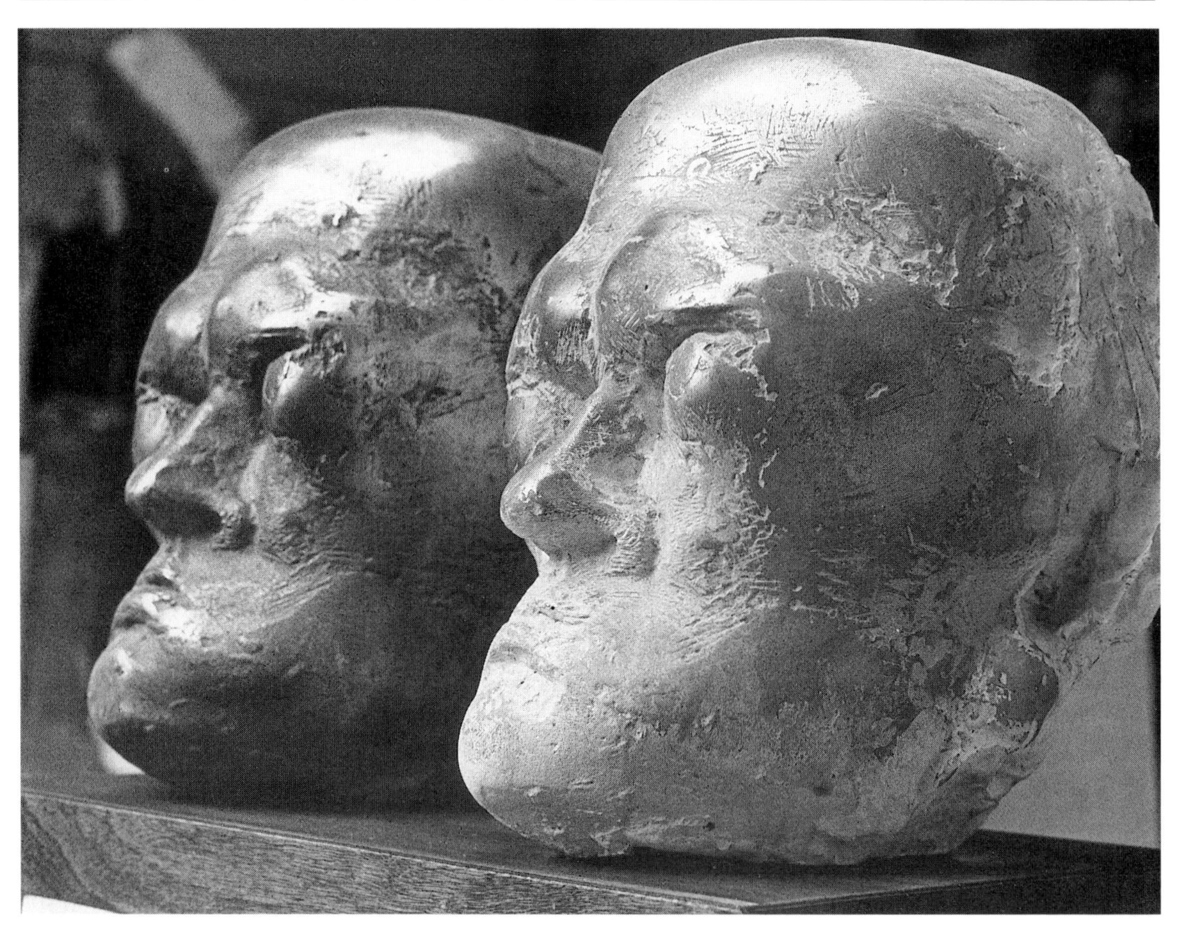

21 and **22** *William Blake* by John W. Mills, sand cast aluminium. The surfaces were defined and finalised on the metal casting.

23 Terracotta doll, 700 BC from Athens. Constructed from pre-cast elements with additional modelling, the elements being produced from a standard mould. *British Museum*

24 Plaster casting of the Trajan column. *Victoria and Albert Museum*

25 Plaster casting of Michelangelo's *David* and other of his sculptures. *Victoria and Albert Museum*

preference over design. Its main virtue is in its unique visual quality and its range of subtle colour and texture.

Other materials require research by the student but occupy relatively recent places in the history of sculpture materials. Napoleon, ravaging the far-flung corners of Europe, looted works of art to be sent home to his beloved Paris. When he came upon such items as the

Trajan Column in Rome, which could not be easily transported he had reproductions made. These were sent either as substitutes for the originals or to convince the exchequer to pay for the removal and shipping of an original. The reproductions were made in what we know as plaster of Paris, and there developed a skilful team of moulders and casters. In the Victoria and Albert Museum, London, are housed some of these same plaster casts, together with many others, numerous, complex and overwhelming in the obvious casting skill which was put into the production of these replicas. These plaster courts are like an encylopaedia of the plaster moulder-casters' art and include castings from the Trajan Column, Michelangelo's carvings, and Donatello's *St John the Baptist.*

Plaster of Paris is not a durable material if exposed to the elements, and is therefore regarded only as an intermediary by most sculptors to be used in the studio or workshop to make the master casts. Indoors it is durable and such things as cornice mouldings and ceiling decorations are cast in plaster of Paris. Adam's ceilings are fine examples of cast plaster decoration. Waste moulds and piece moulds are commonly made of plaster of Paris, also mould retaining cases for flexible moulds of gelatine, polyvinyl chloride compounds, latex or silicone compounds. It is used also in the manufacture of metal casting. The investment or outer refractory mould is made with plaster as the binder, as is the core or inner mould.

Plaster of Paris, together with clay, have been the most familiar materials in sculptors' studios for many years. Today there are many more to be found there. Cements have provided a useful casting material for over fifty years. Some people believe the material used by the Georgians (Coade stone) to have been a form of concrete. It was used extensively for architectural decorations, and can still be seen on many surviving Georgian façades. Coade stone, however, appears to have been of a curious secret

26 *Bather* by Emilo Greco, bronze. *The Tate Gallery*

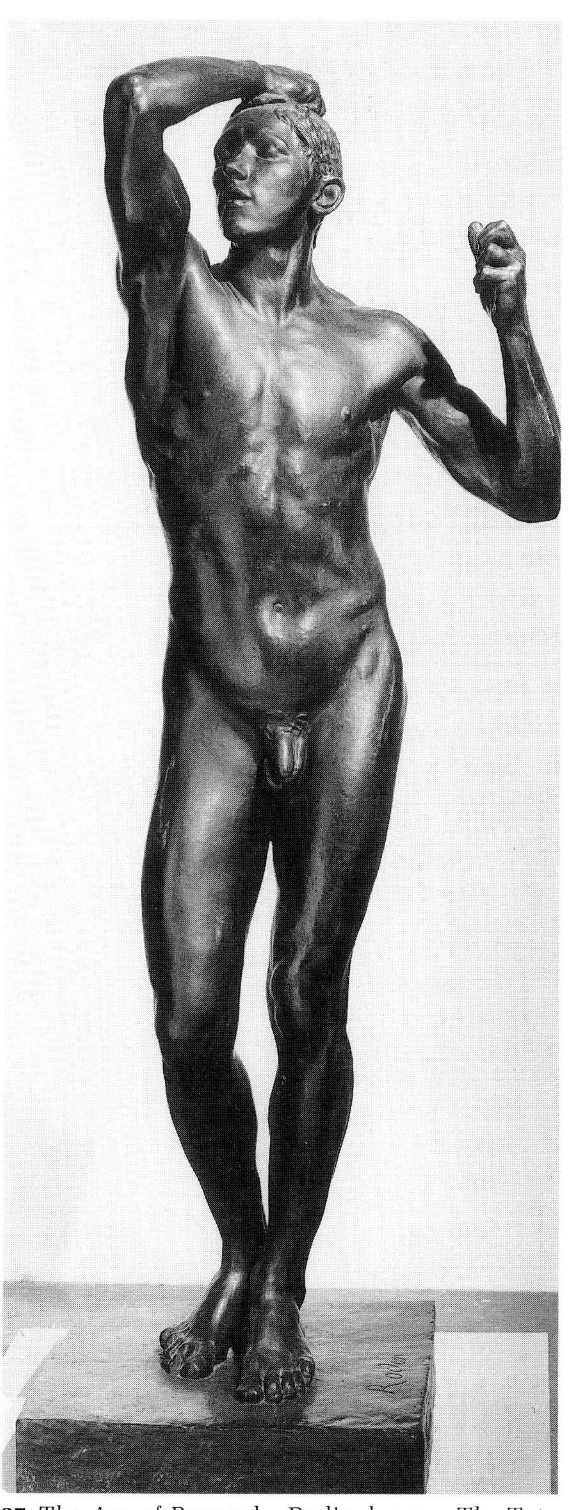

27 *The Age of Bronze* by Rodin, bronze. *The Tate Gallery*

formula, harboured by the particular Coade family who were responsible for manufacturing the applied ornament and taken with them to the grave.

Cements are the binding agent with which aggregates are bound together. The strength of such mixes comes from the aggregate, i.e. the particles or pieces of which are stuck one to another by the cement. Concrete is a mixture of cement and a large aggregate. Mortar is a mixture of a small aggregate with cement. Proportions are important in these mixtures (see page 67). Cements allow the sculptor to produce artificial stone qualities. Henry Moore used reinforced concrete in 1927, casting the basic shape from which he carved the final surface quality and form. I would emphasise that concrete can only be a substitute for stone, never as good as the genuine stone. It is cheaper, however, and allows finer forms to be made, and it is possible to simulate all kinds of natural stone. Early concrete castings were made using the Portland cements, the most common variety, which takes its name from the English stone it resembles in colour. High aluminous cements with their property of hardening more quickly speeded up the whole process of casting. Since its introduction aluminous cement has developed as the principal cement used by sculptors. Qualities inherent in the material have been developed and exploited for their own sake, and I am pleased to say it is not at all common today to see cements made to look like stone or another material. Exposed aggregate castings are a case in point. They are made to exploit the combinations of cement and aggregate as a unique quality.

This aspect of making a new material resemble an old-established one has been the biggest stumbling block in the innovation and exploitation of new materials. Cements were made to simulate stone, to test its quality in comparison with real stone. Plastic materials were made to resemble wood veneers and metals of various kinds. Consequently development was inhibited because it was thought that the new materials had to have the same form as the old. To illustrate this point, I was asked a few years ago to design a body shape for the first of the Lola rear-engined junior racing cars, to be made in cast glass fibre reinforced polyester resin GRP. This presented a most interesting design problem. I examined closely other cars of this material to familiarize myself with particular design-production features. To my dismay I found that although many cars on the race track were of polyester resin, they all affected the shape and structure of beaten aluminium forms. Clearly the problem was to work from first principles to design a cast shape. The final shape included in the casting the instrument panel and various cast elements, body clips, moulding to take the windshield, items that not only cut down on assembly time but added to the rigidity of the shape, making it possible to retain a very lightweight body. It is now accepted practice to cast integral mouldings, but until then this fine new material was made to the same pattern as the old.

Sculptors have also to some extent been remiss in this mistaken simulation. Happily, however, the situation is now much healthier and qualities inherent in particular materials are pounced on and exploited. Shape, colour, idiom and function are all being affected to the good of three-dimensional design generally.

Constructionist sculptors have been most adroit in this, seeing potential in a particular quality and combining this with another; assembling images from diverse material and seeking these in all branches of industry. Casting is, of course, a somewhat different problem. The fact that every material has a set process often results in similarity in forms made in any one substance. Forms are often restricted to a range suitable for casting in the chosen media. The material and the differences between materials are, to my mind, a crucial factor in the development of form and personal expression. These differences must be known, understood and always kept in mind.

Bronze is nothing like concrete, stone is dissimilar to concrete and both are radically different to a resin; resins do not resemble metals when used properly, and these facts are important to a proper marriage between image and material, function and form.

Plaster of Paris is a kind of bastard material, having no parental claim to a particular quality and because of this it may be regarded in the same light as clay and wax. These materials are malleable to a degree that makes them almost characterless. I say 'almost' characterless because clay and wax can be used as a final

28 *Judith* by Marino Marini, 1945, bronze. *Kunsthistoriche Musea, Antwerp*

material. Clay as terracotta (fired clay) has a unique character and should not be made to simulate another medium. Wax too has certain peculiar characteristics that ought to be exploited when it is used as the final medium. As intermediaries, which is their main function today, they are to be controlled and dictated to absolutely. They offer no resistance and are therefore to be moved and massed, and made to take the form of the image in relation to the handling, and the final cast material.

Synthetic resins or polymers have a very wide range of qualities peculiar to particular types of resin. The types of resin are numerous, in industrial application they must number in hundreds, each bearing its own characteristic. They can be used to look like traditional materials, but obviously this is an abuse. Synthetic resins fall into two main categories, thermoplastic and thermosetting. *Thermoplastic* resins are those that become pliable and plastic when heated. *Thermosetting* resins become hard and insoluble by heating. On the whole from the sculptor's point of view thermoplastic resins are used for making moulds and thermosetting materials are used in the main for making castings. Moulds of small forms can be made from polyester resin (thermosetting) and used to make castings in a variety of media. These are essentially piece moulds. Expanded polystyrene (thermosetting) is also an adaptable resin, which is used as a pattern making, or moulding material or as a direct medium. Sculptors concerned with producing large architectural concrete sculptures have found this to be a most useful substance from which complex shutterings can be easily made to make a mould; the shuttering being in fact the equivalent of the sides of a mould, in which detail and texture can be made. This is the female or negative mould into or against which the casting material is poured or packed to make the male or positive form. When the casting has hardened and cured the polystyrene shuttering is removed. This can be done by burning or by painting on styrene, both methods cause the polystyrene to volatilise to reveal the positive form. The casting substance can be any cold setting material, i.e. plaster of Paris, cement, and certain resins.

Expanded polystyrene can be used to make a positive form, which, upon completion, is provided with a pouring gate (a system of risers and runners) and a suitable refractory mould. Molten metal can be poured directly into the mould, on to the polystyrene, which will volatilise to be replaced by the metal, to make an efficient and cheap cast metal form.

Both clay and wax are useful moulding materials. The practice of casting delicate forms by means of wax moulds, which could be easily removed without damaging the cast, was used extensively by Victorian sculptors. They were concerned to avoid damage to delicate forms on a figure caused by vibrations set up when chipping away a hard mould material. Wax moulds are removed by warming the wax to make it pliable, then peeling off the mould with care to reveal the cast. This technique is extremely useful and is an interesting and

29 A cube constructed from wax casting taken from clay negatives by Paul Hands, first year student, St Albans School of Art.

quick method of teaching simple casting processes. Wax used with suitable reinforcing wood, mild steel, etc., can be used to mould very large scale work. I have seen a figure 10 feet (3.1m) high moulded in this way. The original or positive form was built up using an extraordinary mixture of materials including clay. Because of this mixture the sculptor was unable to make the usual plaster waste mould. What he did, therefore, was to clay wash the sculpture and then spray molten wax all over it. He built up a substantial thickness reinforcing this to make it rigid. When the covering was complete he simply cut through the wax opening the mould to remove it from the original and make the casting. The impermeable surface of the wax makes it an admirable mould for plaster, cements, and resins.

Clay is also a very versatile moulding material and is often used to make press-mouldings. Press moulding in this sense meaning pressing clay around an object to make a mould. Sculptors often make such pressings from items such as stones of extraordinary shape, wooden objects, machine parts or other sculptures. Indeed almost anything with a hard surface can be moulded in this way. Casts in plaster, cement, wax or resin can be made from such moulds although plaster of Paris is most commonly used. By cutting, adding and adjusting, a new image can be made from the shape. Henry Moore made great use of this practice, retaining organic strength and form found in a particular shape and incorporating this in his imagery. Complex forms demand a kind of clay piece mould. The separator, used to avoid adhesions of the clay to the object, is usually French chalk. I often use this technique to duplicate a form I have already made and which I want to retain, but at the same time wish to vary the image or use it as the basis for another.

30 Plaster relief made by pressing items into clay, therefore preparing a negative; a plaster cast from this produced a positive. Made by Derek Josey, prediploma student, St Albans School of Art.

The desire to illustrate contemporary sculptures in the newer materials is very great but cannot easily be realised. To say, look at Sardinian or Greek bronzes as examples of modelling directly in wax is one thing; similar technique and form may be studied in Greek and Egyptian bronzes. Nearly every capital city in the world can supply museum examples to fit such a bill, but in the field of present-day sculpture this is not the case. It is vital to study new techniques and forms for such knowledge to have a cumulative effect. Advice to students would be to see everything it is possible to see, in the realms of three-dimensional design and study the material and the technique relative to the form. Materials which are essentially twentieth century are so far relatively little known. Indeed there are many materials, such as cast nylon, which are as yet not tried and their particular qualities as media for sculpture remain as yet undiscovered.

Of modern artists, the man who so often, in so many ways, and in so many materials put a finger on a particular quality to indicate a use, is Picasso. He it was who indicated potentialities in materials with regard to imagery. He tossed into the air ideas and idioms leaving them to be consolidated by others. Picasso made rediscoveries and innovations because of his instinct for character and specific quality in media. By his example many things have been attempted in the twentieth century, some achieved. Out of this activity must surely emerge a twentieth-century character, and materials must play an important part in this, whatever it will be.

The following are specific examples of things I have seen and taken note of in the newer casting media, such as concrete, the various resins, and alloys such as aluminium. In America, Frank Gallo works in polyester resins and makes provocative polychromatic sculptures, an example of which can be seen in the Museum of Modern Art, New York. Also in this museum there is an interesting sculpture by Ernest Trova in cast bronze, which has been chromium plated (*31* and *32*). In England, Eduardo Paolozzi has made recently some chromium-plated sculptures. Concrete sculpture has been widely developed. The late London County Council, under its patronage of the Arts scheme, placed many concrete sculptures in prominent sites in schools and other public places, in and around the greater London area. To my mind one of the most interesting and successful integrations of sculpture and architecture can be seen and enjoyed at Newhaven, Connecticut. Eero

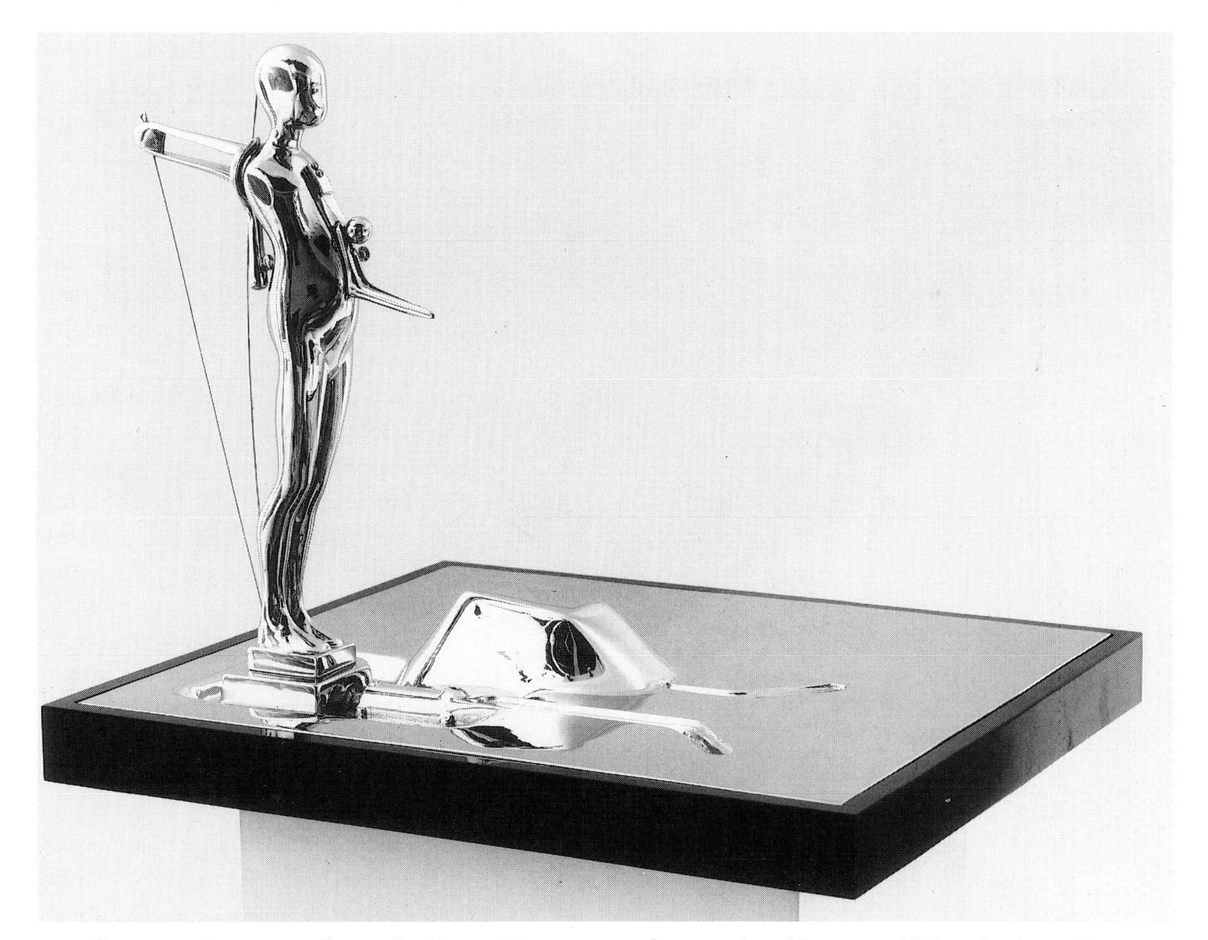

31 *Falling Man Series, Landscape* by Ernest Trova, 1965, chrome-plated bronze with formica base. *Hanover Gallery, London*

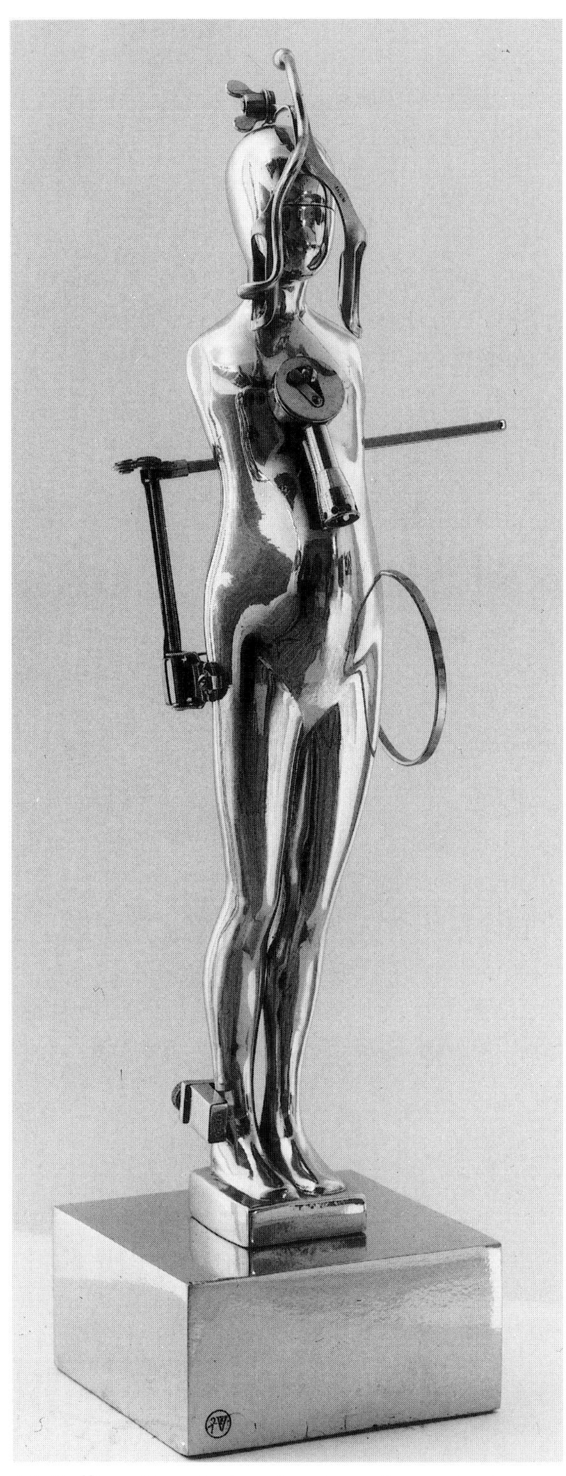

32 *Falling Man Series, Standing Figure* by Ernest Trova, 1965, chrome-plated bronze in transparent case. *Hanover Gallery, London*

Saarinen commissioned Nivola to make sculptures for his residential building at Yale University (*35* and *36*). The result is a great number of concrete sculptures placed in sympathy with the building. In places it is used humbly as decorative features, in others as definite, fully realised, freestanding sculptures, with great strength. The total effect is one of completeness, and walking between the buildings is to experience an ambience that is often felt only in Renaissance Italy, or the natural (not man-made) world. Too often new building, barren of sculpture, gives the feeling of being incomplete, like a salad with no dressing, wholesome but bland. Blank façades stare down empty eyed and dull, begging for something to release them at least from their own state of undress.

Architecturally, concrete has been highly developed. The buildings of Nervi are superb examples of the finesse it is possible to achieve by using pre-stressed or reinforced cast concrete. Japanese architects have taken the art of cast concrete to a very fine degree of grace and accomplishment. Components are cast from moulds, built in situ (shuttering) and prefabricated, cast from moulds made of metal, polyester resin or wood. The concrete is poured, reinforced and consolidated to reproduce exactly the surface of the mould. Some buildings constructed in this way are, in fact, a kind of huge sculpture and the impact of their form has had an effect on sculpture, as they in turn have been influenced by sculptural concepts of space and form.

Plastics are the least known of modern materials, and so relatively little can be said about them as media for sculpture. I hope in fact that young readers of this book will break new ground in this sphere and develop plastics for sculpture in pace with industry. In industry, of course, the application of plastics is enormous and developing every day. The needs of the industrialists create a vast forcing ground for development and research, and the sensible student will take advantage of this

33 *Helmet Head* by Henry Moore, bronze. *The Tate Gallery*

34 Pope Giovanni by Giacomo Manzu, 1963, bronze.

35 and **36** Concrete sculptures by Constantino Nivola. Designed to be incorporated with the architecture of Eero Saarinen in the residential building at Yale University, Newhaven, Connecticut.

fact and utilise the discoveries and knowledge coming from industry.

Machines have been developed for the industrial application of plastics, machines capable of producing very high quality and high fidelity casting. The cost of such machines is, of course, prohibitive for the sculptor to use in the studio, but the possibilities for designing forms and schemes to be produced by such machines is certainly worth investigating. The possibility of casting in nylon for instance becomes an exciting prospect. This material has many fine qualities and has an enormous application in the engineering field that could also be used within the realms of sculpture, and certainly in the realm, generally, of design. On the whole this material is cast using an injection moulding machine. This machine heats the raw material to make it plastic, injects the molten plastic into a mould, under great pressure, then cools the casting and removes the mould. The cost to operate the machine is, of course, very great, and limits its use mainly to production runs. The production of large editions of a work, often in the region of 1000, is becoming more common, although in general, the sculpting profession remains geared to producing unique single works. It is conceivable, however, that a single vast wall relief could be made by producing components to assemble. In fact, the possibilities could be enormous. The variety of colour, shape, strength and form it is possible to achieve indicates an area of activity for sculptors which could prove to be very rewarding.

For the purpose of description of the various methods of making moulds and castings, it is useful to design some basic shapes that allow the various principles to be learned. The human figure comprises the most complex organic form, perhaps too complex for the beginner. Certainly it can confuse basic principles and hinder the proper learning of skills of various kinds, including moulding. I would suggest using simple forms, which can be easily equated with those of the human figure but which do not inhibit by undue concern for such detail as features.

(1) A sphere or enclosed form having obvious connotation with the head. (2) A cylinder = the torso. (3) A cubic form = the pelvic mass. (4) A shape which is composed of two cylindrical forms surmounting another rectangular volume. This can be used with the cylinders surmounting or surmounted by the rectangular volume and can be regarded as equalling legs and base, or legs and pelvic mass. These shapes should be modelled in clay to make the originals to be waste moulded and cast in plaster of Paris, cement and resin. The castings so produced will be the hard originals or master casts, from which piece moulds can be made, and flexible moulds from which more castings can be taken in various materials. In this way, detailed examples are given in this book, and in the studio valuable experience and knowledge achieved, from which a student can make his own progress in a particular sphere.

It is possible to make only general statements regarding casting methods employed on particular sculptures. Demonstrations given by following clearly a specific sculpture through its various stages is more suited to cinematography than a book. Where it is possible, illustrations of simple exercises will be supplemented by illustrations of parts of the particular process on specific sculpture. The skill required to cast with confidence is, of course, considerable and the only way to achieve this is by practice and experience.

Mistakes will inevitably be made, sometimes very serious ones, but the best advice I can give to the caster is that he should think of the problems of a particular casting well ahead, and work out a methodical programme of tackling them. It is foolish to rush in and risk making a mess of things.

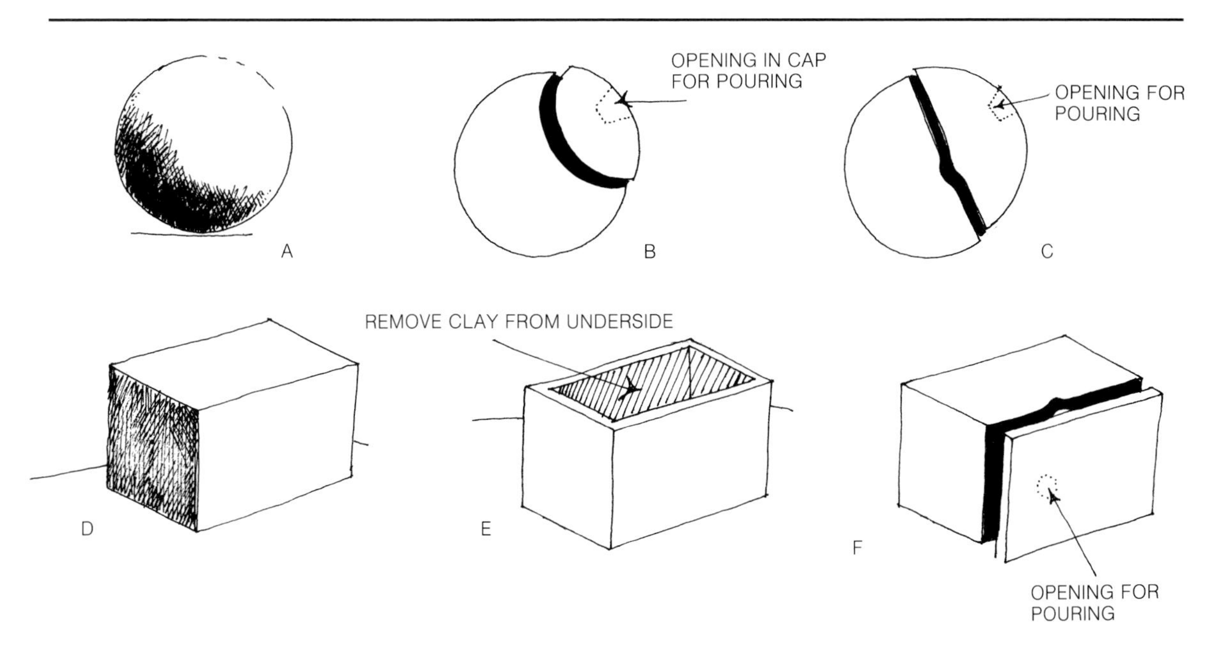

37 Basic forms: sphere and cube. A and D suggest form on which to practise. B,C and F show seam divisions from which clay can be removed simply. E is an open-ended form.

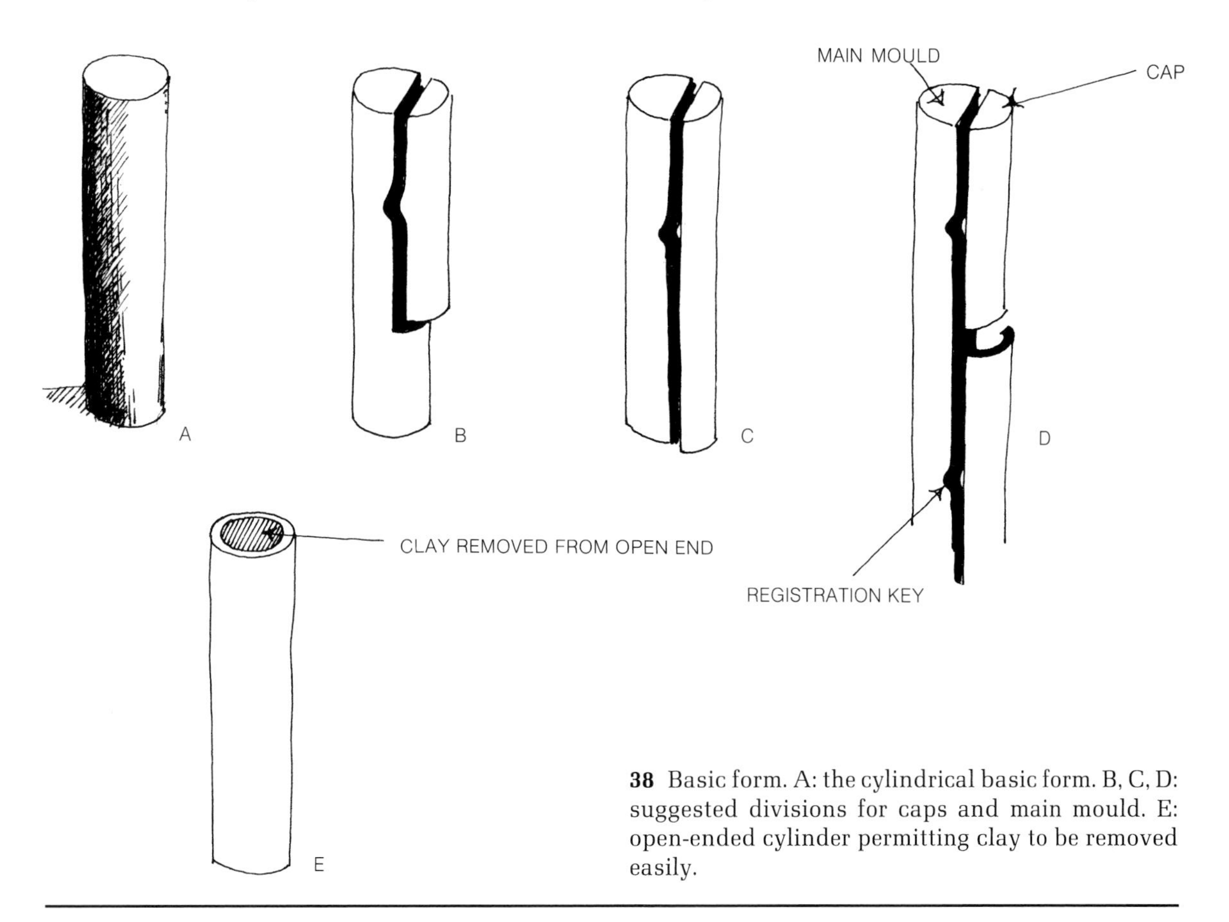

38 Basic form. A: the cylindrical basic form. B, C, D: suggested divisions for caps and main mould. E: open-ended cylinder permitting clay to be removed easily.

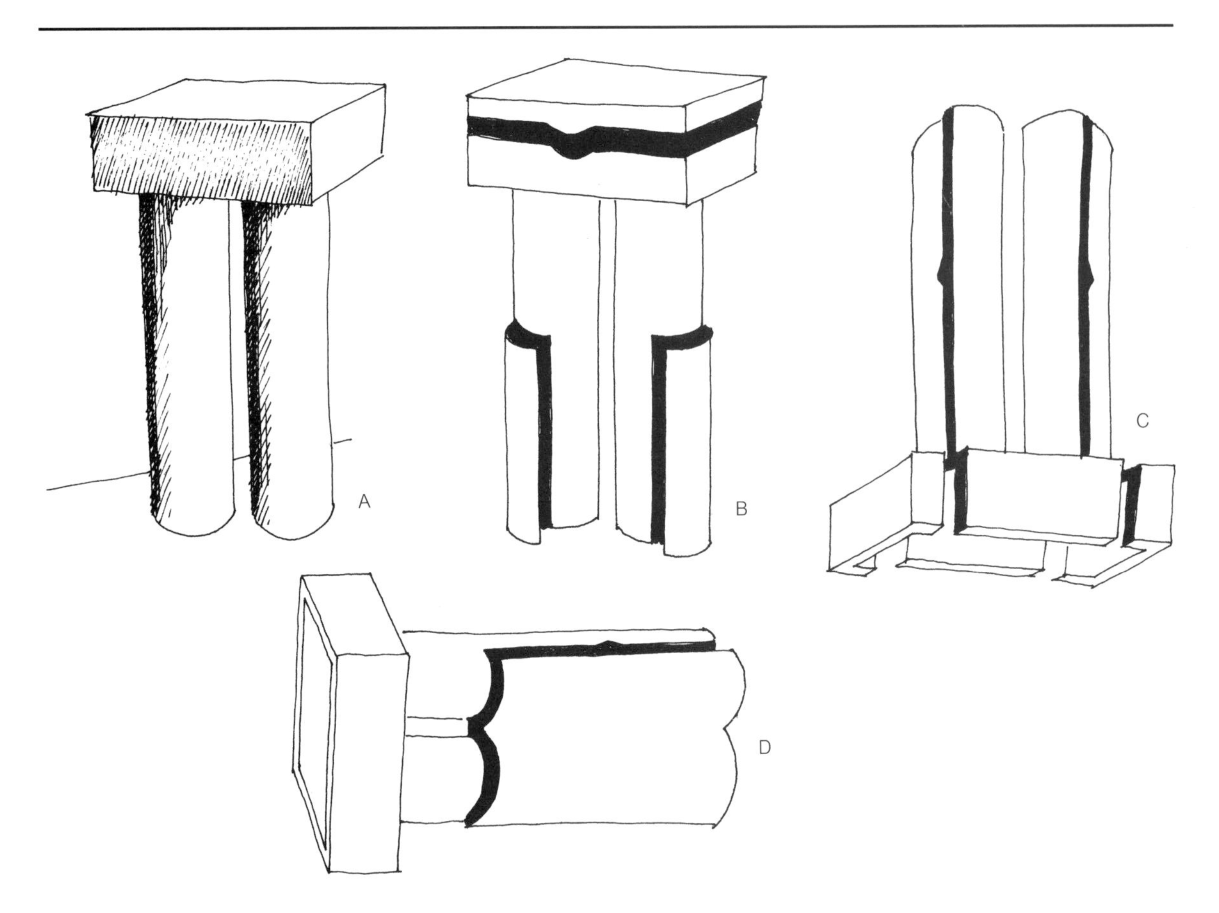

39 Basic form. A: suggested practice form. B,C,D: possible alternatives for design for caps and main mould.

II Plaster of Paris

Plaster of Paris besides clay is the material most relied upon by sculptors. It has many uses, such as fixing stone to stone for carving, holding an armature firmly together, and modelling directly. It is an essential ingredient to all the processes of casting and demands priority of explanation.

The name plaster of Paris derives from the fact that gypsum, from which it is made, was first mined for the commercial development of plaster at Montmartre, near Paris. Now of course it is mined in various other parts of the world. Historically a form of plaster has been known since the Sumerian civilization. It was made by grinding gypsum or chalk and mixing this with ground marble or stone to give it body. This powder was then mixed to a paste with glue size as a binder to make it set and harden. The resulting plaster was used to make rough impressions or built up to make an image. This mixture, or mixtures similar in kind, is known as gesso or stucco. The development of the technique, which led up to it being mined and developed for industry, must be taken for granted. There are accounts of various artists and artisans making a plaster for their own use. Verrochio, for instance, made such a plaster. After use he broke it down and remade it. Giorgio Vasari in his book, *Lives of Seventy of the most Eminent Painters and Sculptors*, tells of Verrochio's technique, 'he made his moulds from a soft stone found in the neighbourhood of Volterra, Sienna' (this must have been alabaster) 'and other parts of Italy, which being burnt in the fire, powdered finely, and kneaded with water, is rendered so soft and smooth, that you may make it into whatever form you please; but afterwards it becomes so close and hard that entire figures may be cast in moulds formed of it.' Verrochio used this material to make castings of natural forms from which he could work, or have as Vasari puts it 'objects as he desired to have continually before his eyes'. This included casts taken from the living model.

The manufacture of plaster of Paris today is extensive, and the process is similar to Verrochio's in that the raw material is ground, heated and in some cases mixed with a binder. The basic raw material is the mineral, gypsum (hydrated calcium sulphate), quarried largely from underground deposits in various parts of the world. It is mined in a solid crystalline form, which is broken down by heating to partially dehydrate and de-crystallise. It is then crushed by steel balls revolving in large cylindrical drums, and ground to produce the plaster we recognise in the studio. When the powdered material is mixed with water it re-crystallises to form a solid, returning virtually to a hard form as in the original gypsum. Re-crystallisation is accompanied by heat, the heat is roughly equal to that used to partially dehydrate the raw material. The question always asked is, can plaster once used be used again? The answer is that it can be, simply by using the process described above, but the effort involved would be too great to justify the result. I would regard this in the same light as manufacturing plaster in the studio from the raw material, the return would be too small for the great effort involved. Industry produces plaster for every conceivable use, from fine casting plaster used by dentists, to very coarse slow setting stuff to use on walls and other building surfaces. Between these extremes, plaster is made to be extra hard or to have great or little expansion on setting. It is made to produce the finest detail or as a foundry pattern with a refractory additive. It is a very adaptable material, and industry will most certainly require a plaster that also suits an individual sculptor's requirement.

The greatest drawback, from the sculptor's point of view, is plaster's inability to stand up to the weather. Many methods have been tried to make it weatherproof. It can be immersed in boiling linseed oil, which is the most successful method, but this is made difficult by the size and capacity of the vessel required to achieve complete immersion. A surface coating resin can be applied, but if this does not completely cover every surface perfectly, moisture will penetrate and get under the resin and cause it to peel. It can be treated with such a resin or shellac or liquid wax for a short-term life in the open air without deterioration. Frost, of course, will make such a life even shorter.

This chapter is concerned with the use of plaster of Paris principally as a material for making moulds and taking casts. Its application in this respect is many and varied. There are two main types of plaster moulds (1) waste mould and (2) piece moulds.

(1) *Waste Mould* The most common plaster mould is known as a waste mould. The name derives from the fact of the mould being broken from the casting and made waste. This kind of mould is used generally to make one cast from an original soft material, such as clay, wax, Plasticine or Plastalina. Waste moulds are designed specifically to facilitate a speedy production of one cast, usually the master cast. Castings can be taken from such moulds in a variety of substances such as plaster of Paris, cements, polyester resins and hard wax. The mould has to be designed to allow the particular material to cast satisfactorily and to allow the removal of the original clay.

(2) *Piece Moulds* As the name suggests a piece mould is composed of a number of pieces. Each piece of the mould is designed to allow every undercut form a number of parts to enable it to be cast and removed easily. It is a method of moulding used to make more than one cast, the number of castings being dictated by the nature of the material to be cast. The pieces are held by a case or jacket of plaster. This ensures positive register of the pieces in relation to each other, and consistent accurate casting. This type of mould can be taken from hard or soft originals but usually from a hard substance. It is possible by this method to make a mould from virtually any completed sculpture. Plaster of Paris, cements, clay, wax or resins can be cast from a well-made piece mould.

Mixing Plaster Plaster of Paris is mixed with water and re-crystallises to form a solid. It should be mixed in equal volumes of plaster and clean water. The process of change is from a dense opaque liquid to a more dense state, gradually becoming stiffer in consistency till it solidifies and hardens. The action of re-crystallisation, whereby the plaster returns to its original hard form, is fairly rapid and causes the plaster mass to become warm. This heat is roughly equal to that exerted upon the raw material during manufacture. The plaster will not be strong until after this heat has cooled. It becomes harder as the mass cools and dehydrates. Do not disturb the plaster until it has cooled.

Most sculptors use a rough guide to proportions of a plaster mix. The measure of water should be half the total quantity of mixed plaster required. The plaster should be stored in a dry atmosphere to prevent it deteriorating by taking moisture from the air, as it is a very hygroscopic material. Before adding the plaster from the bin to the water, stir it to aerate it as this facilitates quick and even mixing. Then with the hand add the plaster to the water. The mix is best made in a plastic vessel that is flexible enough for the left-over material to be easily removed when hard. Add the plaster to the water allowing it to run through the fingers. Do this until the plaster is just covered by the surface of the water, and a small mound of it appears in the middle of the vessel above the water. Add one extra handful to this (I always do), then leave to stand for a few seconds. It is possible to leave some plasters in the water at this point for as long as 10 to 15 minutes. It does not begin to set properly until it is stirred and mixed thoroughly. After leaving it to stand, mix it thoroughly with the hand, until it is free of all lumps and is of an even consistency. The plaster may cause irritation to a sensitive skin, in which case use a barrier cream and mix the plaster with a spoon or some such implement, and prevent the plaster from setting on the skin. The prepared mix should look opaque on the hand; if it is not, it is too thin and will be a very weak plaster. It is advisable to experiment with small mixes to become accustomed to the look and feel of a good mix. Never make a mix of large quantity unless you know exactly what you are going to do with it. Always make an experimental mix from a plaster you are not familiar with. Mix small amounts that can be handled easily, rather than large quantities which cause you to hurry and make mistakes. Try to mix plaster consistently throughout a work to control expansion and contraction rates. Uncontrolled and uneven mixes will result in cracking and sometimes lack of adhesion between applications.

The rate of setting for any plaster can be altered to some degree. It can be retarded allowing more time to gain greater control. It may be accelerated when a quicker set is needed. To vary the setting time of the plaster it is necessary to include in the mix certain additives.

To accelerate the set, mix the plaster with hot water. This speeds the set effectively and is least harmful to the final plaster. Salt can be added to cold water to accelerate the plaster setting. This is perhaps the most common method used, but the chemical alum, if added to the water, will increase the speed of setting. Both salt and alum should be added in the proportion of one teaspoonful to two pints (5ml to 1.16) of water and should be allowed to dissolve before adding the plaster. The more salt added the quicker the set.

Most modern plasters can be bought with a setting speed selected according to the job in hand. It is not often that a sculptor needs to accelerate a plaster mix. Too often the setting time is too fast, a fact often proved by the large quantities of set plaster in buckets in bowls abandoned in art school studios.

Retarding agents are therefore more often in demand. The most common of these is ordinary glue size. When added to the water this not only retards the setting time, according to the density of glue size, but also increases somewhat the strength of the final material. Very cold water retards the set, as most sculptors will testify, and winter conditions often hinder work in this way, as well as in many others. Alcohol added to the water will also retard setting, as will the addition of acetic acid. The latter should be used in very small proportions. Old casting workshops used the simple additive of urine.

Waste Moulding

This method of moulding is probably the most widely used. It is one of the quickest, and certainly the most direct, moulding technique. The title is, in fact, a description of the technique, the mould being broken from the cast, becoming waste. The mould consists of the main mould, that part containing the greater portion of the sculpture, and the caps.

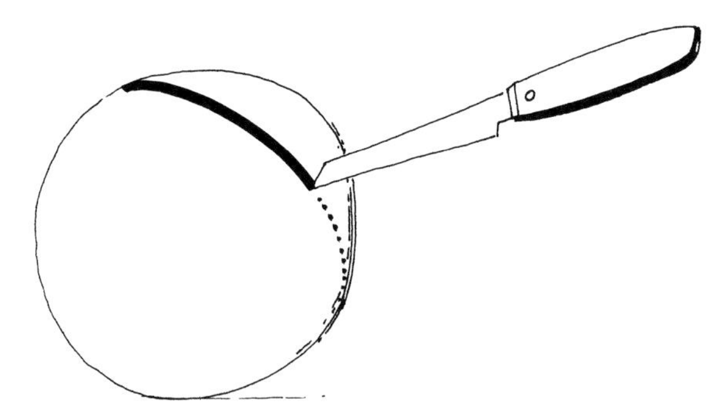

40 The seam line drawn with a knife before placing the brass shim. It helps to plan the main mould and caps.

The caps are removable parts of the mould which, when taken off, allow the modelling media and armature to be taken out, and replaced by the casting substance, plus any reinforcement that is necessary.

Model several spheres, roughly the equivalent in volume to the human adult head. These should be modelled from clay, Plasticine or Plastalina. The modelling material should be kept fairly soft. Clay should be free from any sand or grog.

Draw a line around each sphere with a knife (*40*). This line indicates the seam or division between the main mould and cap of the mould. The line is drawn to make a guide, to assist the design of the mould. The design, which determines the number and size of caps, in relation to the main mould, is made according to the form and the material from which the casting will be made. For instance, the materials which can be poured as a liquid into a mould, then left to set and harden, require fewer caps than the materials, which demand access to the total mould surface, such as concrete, wax or polyester resins. Forms that are bulky and large in volume can often be moulded with few caps, whilst slender complex forms usually require a great many. The proper design of a mould, taking into consideration the relevant facts, is a skill that can only be achieved with practice.

The methods of making the divisions of a mould to make the caps and main mould are as follows:

Brass Shim 5-thousandth in. (0.1mm) brass sheet, known to sculptors as brass shim or fencing, is most commonly used to effect the divisions of the mould. Strips of this material are inserted along the line drawn on the form (*41*). The strips should be approximately 2in. (51mm) long, and after being pushed well into the clay protrude about ½in. to ¾in. (13mm to 19mm). Of course, the strips are tailored with scissors to fit the form, paying special attention to corners and sharply turning, intricate shapes. The line of shim should be even and without gaps, the top edges continuous, and without protruding sharp edges to cut the fingers. The reason for pushing the shim well into the clay is to make it firm and able to withstand the weight of plaster when the mould is being made. Too often I have seen delicately placed shims knocked out when the plaster of Paris was being placed. The shim remains in the mould until the caps are removed.

Clay Wall Another method of making the division between main mould and caps is by using a clay wall. This method is used mainly when no brass sheet is available, or when the modelling material is too hard or too coarse to permit shims to be inserted. It is used also when the flash at the dividing line, on the cast, must be minimal. It is also used to make the caps of a piece mould, which have to be made separately. Strips of clay are placed along the drawn line in such a way as to provide a clean edge on one side of that line (*37*), the opposite side of the clay strip being supported by small buttresses of clay. Plaster is placed against the clean surface of the wall to make one cap. The clay is removed when each cap is hardened, the edge trimmed and keyed and treated with a release agent to make it separate from its neighbour or from the main mould.

Thread This is another method of making the division. It is a hazardous but quite exciting technique and is sometimes used for making a mould from the live model. The technique is to place strong thread or wire along the line of division. The thread is placed directly to the clay and pressed to remain in place. On surfaces that are not moist the thread can be made to stay in position with Vaseline (petroleum jelly), grease or tallow. These can also be used as the release agent on the rest of the surface. The ends of the thread should extend and be held clear when the plaster is placed. The thread is pulled up through the thickness of plaster quickly, just as the plaster hardens. The hazard is in catching the plaster

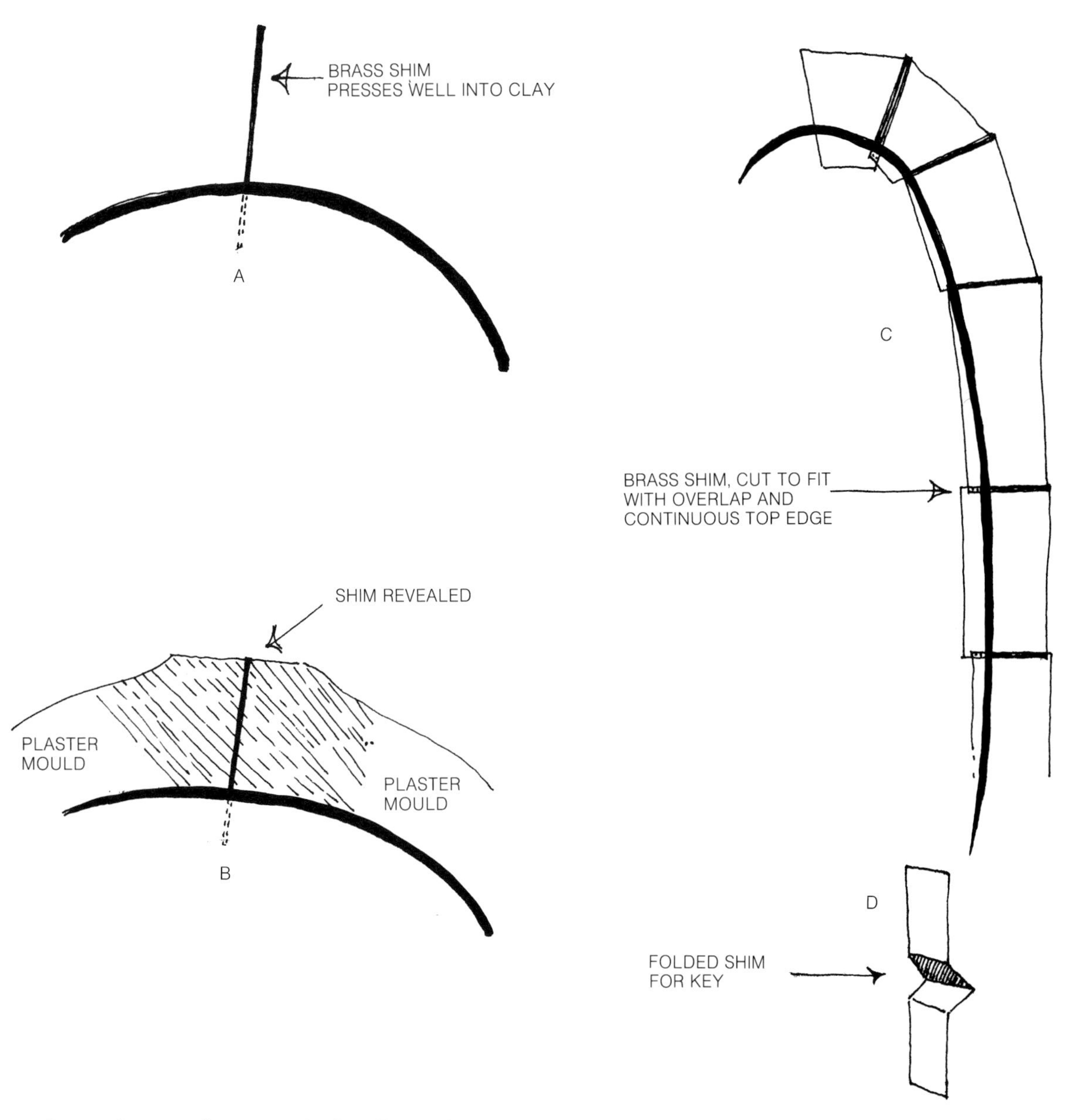

41 Brass shim. A: shim is pushed well into the clay. B: plaster is placed well against both sides of shim, with the top edge revealed. C: shim is cut to fit the form and made to overlap, with a continuous top edge.

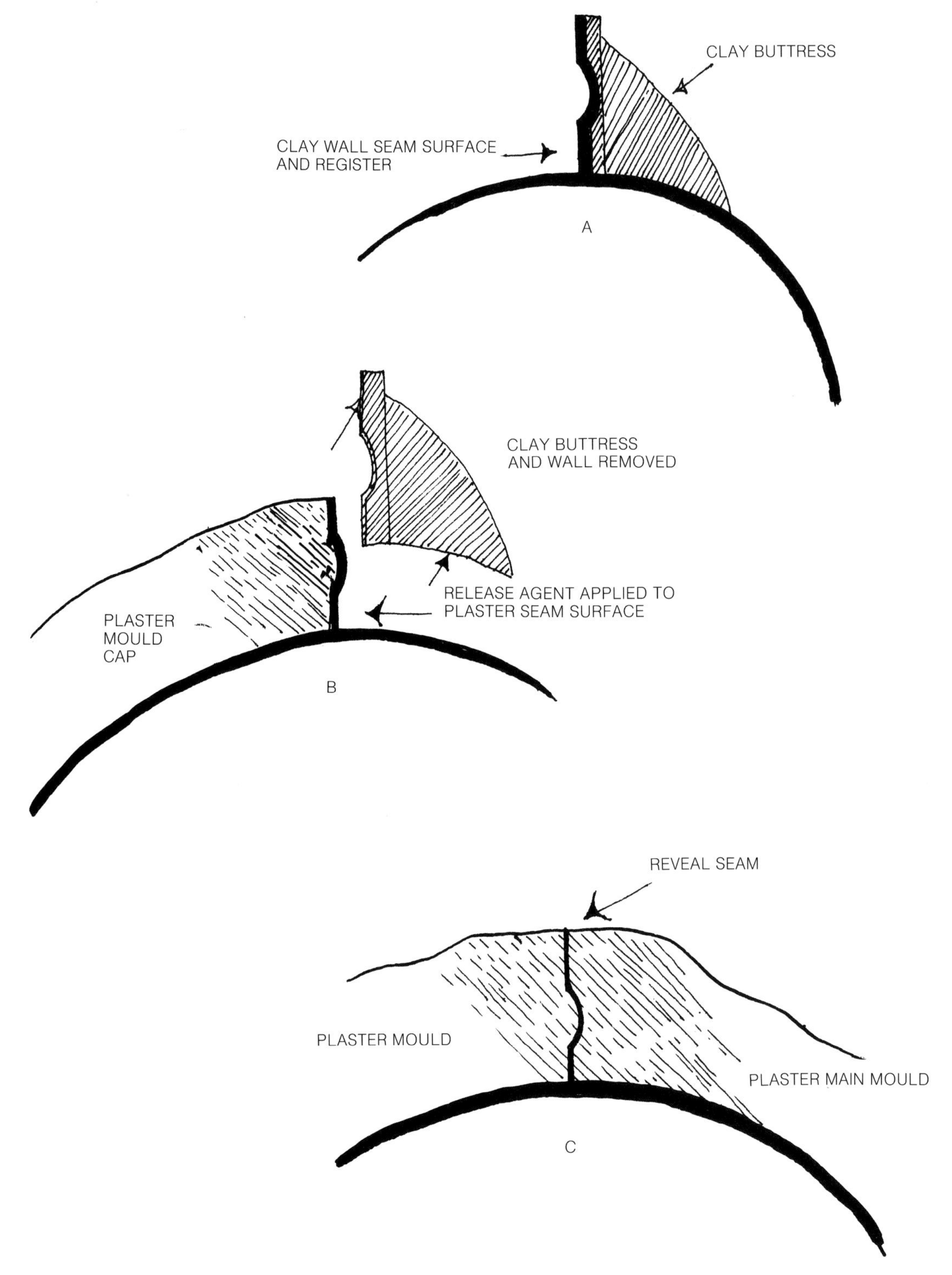

42 Sections through clay wall. A: clay wall and buttress. B: first cap made – clay wall removed and release agent applied. C: main mould made and seam revealed.

at precisely the correct moment. If the thread is pulled too soon, the plaster will not divide. If it is pulled too late, the plaster will be already too hard for the thread to pull through at all. This technique can be very effective but requires practice. Many is the poor live model who has been encased in hard plaster of Paris.

Sawing A crude method of dividing a mould is simply to saw through the thickness of plaster placed over the modelling. This is a particularly useful technique for dealing with a large simple volume. I often use the method to introduce beginners to the techniques of moulding. Although a crude method it is often expedient; the loss of plaster at the saw cut being fairly slight.

No matter what method has been used to make the divisions, seam edges must be continuous, even, clean and unbroken, and, most important, the angle of the faces between the caps must avoid trapping one cap to another. The angle should permit the cap to be drawn away from the main mould, or in a particular order one cap from another. This is determined largely by the angle of seam faces, which should slope outwards. Make a number of basic shapes.

To make the moulds using brass shim:

1 Place the brass shim along the drawn line. Make it even and without gaps, with a continuous top edge which avoids corners and has no sharp edges protruding. The shims sloping outwards all round. Make keys in the shim by folding (*41D*).
2 Mix sufficient plaster of Paris for the first coat. It is helpful if this first coat is coloured so that it will serve to give an indication of where the cast is located when the mould is being chipped off. The colour can be made by adding to the water a pigment or dye, such as yellow ochre or cochineal. It is wise to test the effect of any colouring used on the plaster before making a large mix. Some dyes may kill the plaster, which will then never harden.
3 Cover the form, clay and shim, with the first layer of coloured plaster. Do this by gently, but firmly, throwing the plaster evenly over the model. Throw from a small bowl by dipping cupped fingers into the plaster (half finger length only is necessary). Throw firmly. If you throw too hard, the plaster will splash back at you; if you throw too softly, the plaster will not cover the work well and the mould will be full of air bubbles. Practice will enable you to know the strength of the throw. Make sure the plaster is roughened after each application to make a key for the next layer. Never leave a smooth surface between layers as this can cause flaking on the surface of the mould.
4 After each application of plaster scrape the top edges of the brass shim. Do not allow the seam to be lost under the plaster.

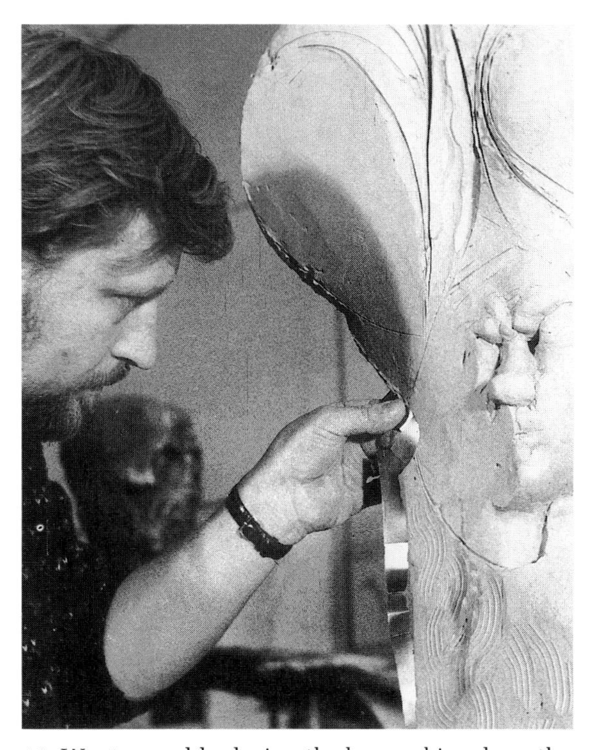

43 Waste mould: placing the brass shim along the marked division in the clay original.

44 Waste mould: mixing the plaster and sieving it through the fingers.

45 Waste mould: mixing by hand to prevent lumps.

Searching and cutting away waste both time and material.

5 Mix more plaster, this time without colouring. Build up with this to the final thickness of the mould. The brass shim will act as a datum in building an even thickness of mould. Build the even thickness making sure of strength at the seam edges. Scrape the top edges of shim to determine this. Some moulds require mild steel strengthening, and this should be placed while you are building up the mould.
6 Allow the plaster to harden for a minimum of one hour, preferably longer.
7 When the plaster has hardened, pour water over the mould or immerse it in water. Insert a knife into the seam and very gently prise off the cap, pouring water to help break the suction between mould and model. Prise from two or more places around the seam. If there is more than one cap, remove them in particular sequence. Lean the caps against a bench or wall. Do not lie them down as they are liable to warp.
8 When the cap, or caps, are off, take away the brass fencing and carefully remove the clay, Plasticine or Plastalina and armature.
9 Clean the mould.
10 Prepare the mould with a release agent, according to the material to be cast. (See the list of release (parting) agents on page 52.)

The technique for making moulds by means of clay wall is as follows:

1 Make flattened strips of clay about ½ in. to ¾ in. (12 mm to 18 mm) in width. This is best done by rolling or pressing the clay out on a flat surface which has been dusted with French chalk.
2 Prepare to make the first cap by placing the strips of clay (walls), with the flat clean side along the drawn line, sloping outwards all round. Fix the clay supports to keep the wall in place, as buttresses on the opposite

side to the line. Press the clay wall firmly down on the modelled form.

3 Mix the plaster for the first coat and colour this with pigment or dye to make the warning coat. Mix enough for the cap only, or for the first cap, if there is more than one. Make sure the surface of the first application is roughened to allow a proper key with the next layer.

4 Mix enough uncoloured plaster to make the complete thickness of the mould on the cap. If it is a large cap, place any mild steel reinforcement required. Trim the plaster at the clay wall.

5 When this has hardened, remove the clay wall and supports. Clean up any marks made on the modelling. Carefully trim the seam face and make keys to allow register.

6 Paint a parting agent on the seam face, i.e. clay wash or oil.

7 Complete all caps in this way. Alternatively, complete the main mould in this way and then make the caps separately.

8 Mix and apply the first coloured layer of plaster for the main mould. Be certain not to lose the seams. Scrape the seams after every application of plaster.

9 Mix and apply the plaster to make the final thickness of the main mould. Place any reinforcement required. Make sure the seams are located and revealed on all caps. Allow the mould to harden.

10 Remove the cap, or caps in a particular order, prising with a knife at the seam. Pour water over the mould before and during this operation or immerse the mould in

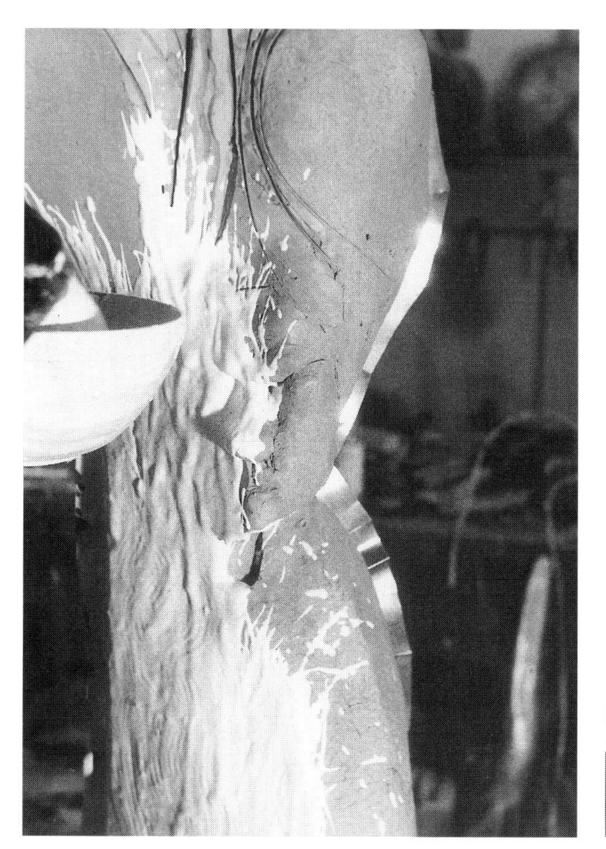

46 Applying the first layer of plaster by gently throwing it onto the clay.

47 Scraping to reveal the brass shim at the seam. This is done after each application.

48 and **49** Reinforcement is placed as the plaster is built up. In this case a complex of steel tubes is being used to make a strong cradle to facilitate moving the mould for filling.

water to release the suction between mould and model.

11 Remove the clay and armature from inside the mould. Clean the mould thoroughly.

12 Apply a release (parting) agent according to the filling to be used.

The mould is now ready for filling.

Make a number of moulds from clay forms which are not cleaned thoroughly after removing the clay. There will remain on the mould surface a very fine film of clay. This makes the perfect parting agent for concrete. Large lumps of clay can be taken off the mould with a soft clay dabber.

The techniques of both sawing through a

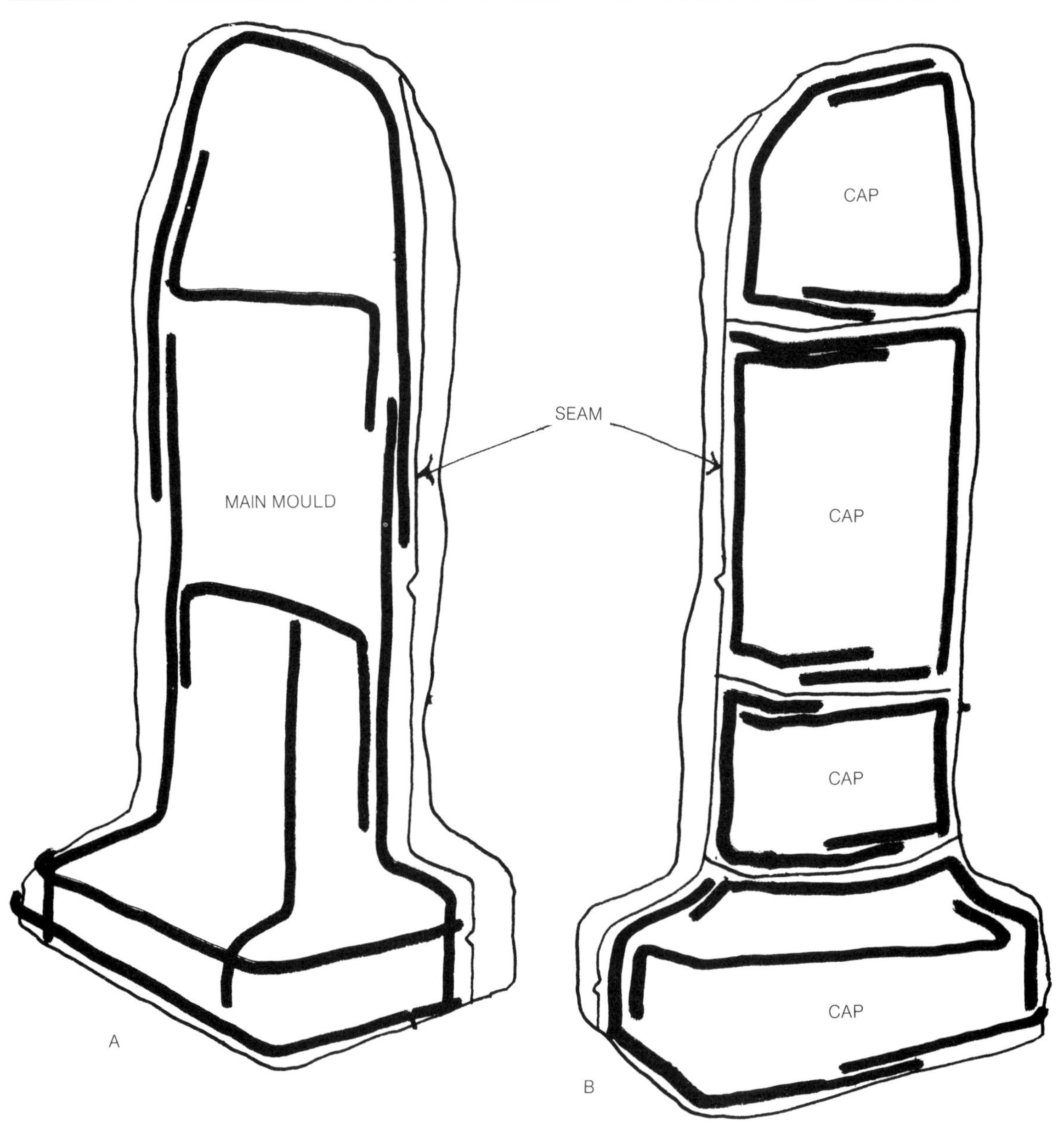

50 Mild steel reinforcing of main mould and caps. A: strength is given all round the mould, vertically and horizontally. B: the same reinforcement is given to each cap.

mould and using a thread can be practised at a later date. The first is an obvious process. The second can be learned as the action of plaster when setting becomes more familiar.

The methods and procedures given for making moulds with brass shim, and clay wall, can be used to mould from all the basic forms.

It is impossible to illustrate or forecast what particular problems might beset particular sculptors. Moulds are as complex, or as simple, as the forms from which they are made. The rule on the whole is to take as few caps from the main mould as possible. Take only those necessary to allow access to the inside surfaces of the mould to facilitate filling with the selected filler.

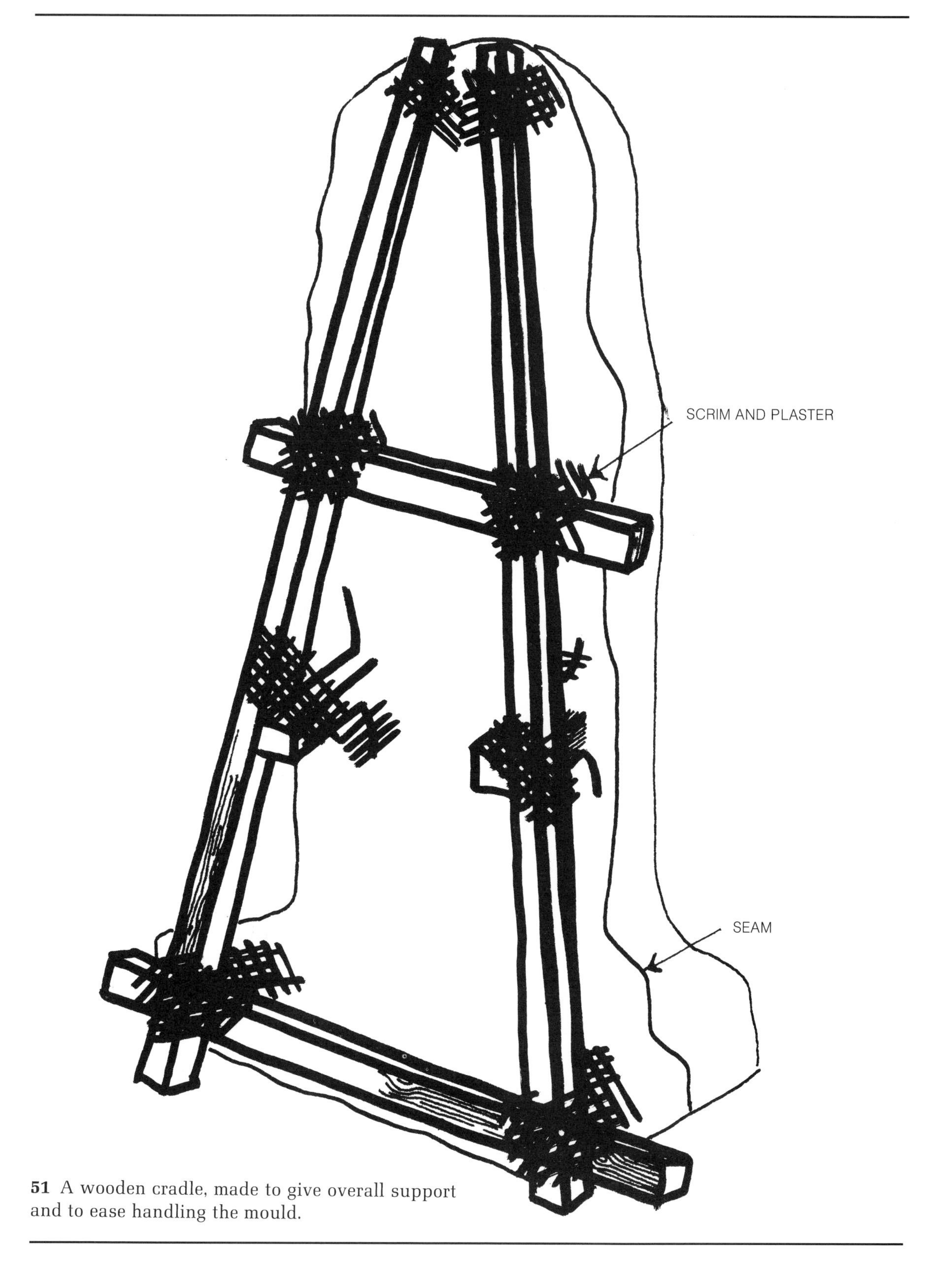

51 A wooden cradle, made to give overall support and to ease handling the mould.

Reinforcement Moulds require reinforcement, and the rule to go by when making them is to make sure that reinforcement is placed to strengthen the caps and main mould. Mild steel should be placed around the seams of the caps and main mould. This strengthening will effect a good peripheral strength if placed properly. Further reinforcement should be placed horizontally and laterally (*50*). It is wise when making a mould of about life size and over, to make a cradle of timber to be fixed to the main mould (*51*). This cradle will give greater rigidity to the mould and make it easier to handle.

Piece Moulding

This technique is the most exacting. Before the development of flexible moulding materials it was almost the only method for producing a faithful mould of a hard original, from which more than one casting could be taken. The principle is to make a separate piece of mould for each intricate and undercut part of the form. The result is a kind of mould jigsaw contained in a case. The pieces fit into the case in a particular sequence and are held firmly in their correct positions. From the mould properly treated the cast is made. The mould is removed by first taking away the case, and then the pieces are taken off in their sequence and carefully replaced in the mould case. It is a most complex technique to describe because everything is so dependent on the nature of the form from which the mould is made. I will explain the basic principle in detail and leave the reader to carry out his own experiments. Working on the forms designed specifically for explanation in this book, piece moulding is relatively simple. Complications occur when forms are more devious and complex, involving minor forms and surface texture. There is an immense amount of satisfaction to be gained from making a good piece mould, however, more so than any other kind of mould. Piece moulds often becoming objects of beauty in their own right.

The basic forms on which to practise piece moulding must already be made of a hard material such as plaster of Paris, concrete, resin, terracotta, wood, stone or metal. If the material is porous, the surface will need to be sealed. This can be done simply by painting the surface with dilute shellac and methylated spirit. Other surface sealers can be used according to the nature of the surface, for instance, polyurethane varnish or a good oil paint or wax. The amount of detail on the original and the complexity of surface forms must be seriously considered and taken into account when selecting the best and most expedient sealer.

It is possible to make a piece mould from a clay original, but it is very difficult to be exact, and piece moulds must be as precise as possible. Precision is more easily achieved working from a hard original.

The sphere is the most useful form on which to demonstrate the principle of piece moulding. It is possible to mould a sphere by making two pieces of mould. This requires absolute precision at the division of the mould. Two perfect half spheres of mould are necessary to allow the cast to be withdrawn cleanly. This kind of precision is not impossible but would take a great deal of time and trouble – probably more time than a sculptor is prepared to spend. If there is a miscalculation in dividing the mould, one or both parts of the mould will lock. This will be frustrating and may cause some damage to the original. To allay any such mishap the moulder makes three pieces of mould instead of just two.

To make a piece mould:

1 Prepare the original or master cast, seal the surface if necessary and apply a release agent, i.e. tallow, oil or a mixture of oil and meat dripping.
2 Draw on the original the dividing lines, then place the original down on a flat

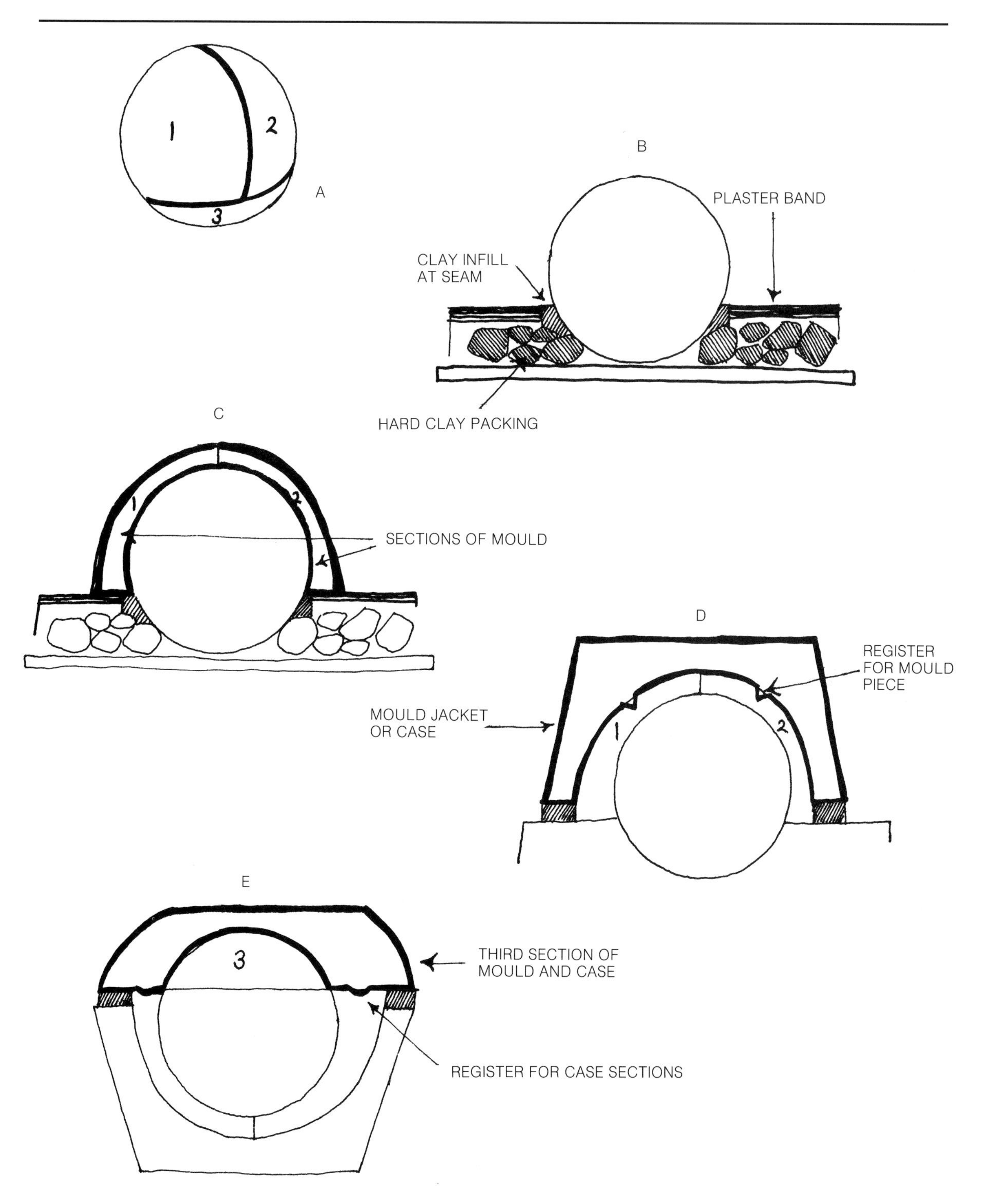

52 Piece moulding. A: describes the piece moulding principle, showing mould sections. B,C,D: the preparation and build-up of the first sections of mould on the original against the plaster band, plus the mould jacket. E: the manufacture of the remainder.

surface, a board or table top, in preparation for making the first piece of mould (*52B*). Prop it up so that it does not move about.

3 Place around the sphere a wide band of clay, the surface of which should be $\frac{1}{4}$ in to $\frac{1}{2}$ in (6mm to 13mm) below the dividing line. On top of this build a band of plaster of Paris, to be level with the dividing line. Smooth the top surface of plaster band. Leave a small space between it and the master cast.

4 Apply some release agent to the plaster band surface.

5 Fill the small space between plaster band and original with clay. Be as precise and clean as possible at the surface of the original. This will determine the seam line.

6 Make a clay wall along the line at the division of the mould.

7 Mix enough plaster to make the first cap, between the plaster band and the clay wall.

8 When this has set, trim the back surface of the cap. This should be smooth. Remove the clay wall and trim the seam edge. Make suitable locating keys in the seam edge.

9 Apply release agent to the new seam edge.

10 Make the second cap. Trim it when the plaster of Paris has hardened.

11 Cut into the back surface of the caps, long registering keys. Apply a release agent all over.

12 Place a clay band along the top edge of the plaster of Paris band.

13 Make the plaster case to contain the mould pieces. Allow this to harden.

14 Turn the whole work over. Remove clay and plaster bands.

15 Trim if necessary the seam edges of the mould pieces and the edge of the case. Apply release agent to these edges.

16 Make the remaining cap. This cap overlaps the edges of the first two to make a grip to remove this from the cast.

17 When the final cap has hardened, remove caps and case from the master cast. Check the mould for any defects, repair these and prepare the mould for whatever filling is to be used.

Make sure that each cap can be removed from the original to be trimmed and cleaned, then replaced. On more complex forms, each piece of mould must be removed to be certain that it can be removed and does not lock.

This is the principle of piece moulding. Protrusions from the surface, forms or texture must each receive a piece of mould. Each undercut must be catered for. Some moulds may eventually consist of dozens of pieces.

The point to remember is that an undercut is the portion of a form that will prevent the mould being withdrawn from its surface. Always double check such hazards, make two pieces and not one, whenever you are not certain. Key the pieces properly to fit one to another and to fit into the case. Make sure the case draws easily from the pieces. Be certain, by numbering, in which order the pieces are removed from the original and the cast.

Filling

The two basic methods of mould making with plaster of Paris enable the sculptor to produce castings in a number of different materials. According to the treatment of the plaster mould in preparation for filling, it is possible to make casts using plaster of Paris, cements of various kinds, terra-cotta, clay slip, and wax, both as part of the process of producing a metal casting, and as a final image. Synthetic resins can be cast successfully using plaster of Paris moulds. Of these materials two, clay and synthetic resins, require an absolutely dry mould, others are cast in a mould that must be very wet.

The first concern when casting, or planning the process of castings, is to prepare a mould which is ideally suited to the substance with which it is to be filled. Some materials can be

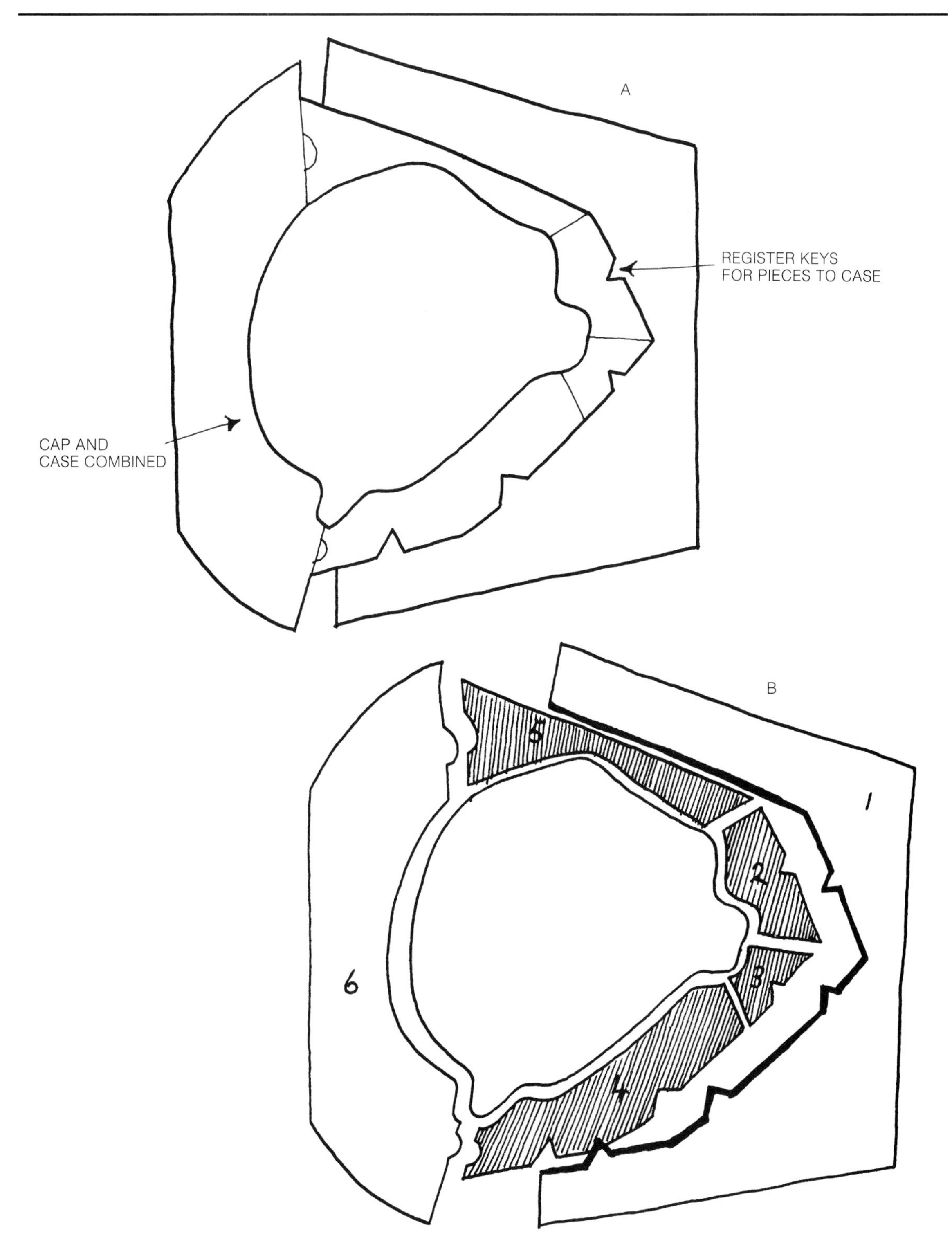

53 Piece mould: diagram of a more complex piece mould and jacket. The exploded plan describes the order of piece removal from the positive form.

poured into a mould in a fairly liquid state, to set and harden. When this is the case the number of caps made for a waste mould can be kept to a minimum. Only enough pieces to allow the removal of the modelling media and armature, proper cleansing of the mould surface, the placing of adequate reinforcing and the pouring of the casting material, are necessary. Plaster of Paris can be poured very easily, and in some instances cements, resins and wax can be poured. With these latter materials special care must be taken to allow for the various properties, and therefore complications, of the media. Such obvious things as shrinkage and expansion of materials during chemical change can cause great damage or distortion if not considered when designing the mould. Polyurethane for instance expands during its chemical change so that a cubic inch (25mm^3) becomes 2 cubic feet (600mm^3) exerting great pressure as it does so. Therefore the mould to contain it must be strong enough to accommodate this aspect of the material. Other factors will become apparent when dealing with specific materials.

Having made the mould with the cast in mind, and having given careful consideration to the casting material, the mould must then be prepared to receive the filling media. This preparation usually entails the application of a release agent, and soaking or drying out according to the filler.

Release or Parting Agents

These are solutions applied to the mould surface to permit the mould and cast to separate easily. They vary in nature according to the chemical properties of the filler. The following list covers the release agents generally used for the various fillers dealt with in this book:

Soft Soap or liquid detergent can be used as parting agents for plaster, cast from a plaster waste mould. Dilute the agent in a small quantity of warm water, to make a strong solution. Apply this solution to the mould surface making a good lather. Leave the lather on the surface for 10 to 20 minutes. The lather is then removed from the surface, using a clean brush and water. When this is done the mould is ready to be soaked and filled. Some sculptors repeat the soaping process and some apply a minute amount of thin oil to the soaped surface. The parting agent seals the pores of the plaster mould, oil ensures this and ensures also a water repellant surface. Soaking the mould is to make sure than no moisture is taken from fresh plaster during filling. If this happens, the cast will be weak and liable to break easily. The surface of the casting will also be defective.

Clay as film left in the mould is a most efficient parting agent for cement, concrete and polyurethane from a plaster of Paris mould. The clay film remains on the mould when the clay is removed. This mould should not be washed clean and any large particles of clay can be picked off the mould surface, using a dabber of soft clay. What remains is the very fine film, which helps a good release but does not blur the modelled detail of the form and texture. If a mould has been well cleaned out, a clay wash can be applied to the surface to make the release. A mixture of clay and soft soap will also prove an effective parting agent. Use this mixture in the same way as soft soap or liquid detergent. The mould needs to be well soaked.

Shellac and Wax seals the mould surface completely, and can be used for almost any filling, particularly for concrete when a minimum of plaster bloom on the cast surface is required. It is effective too for polyester resin casting from a plaster of Paris mould. Apply dilute shellac-methylated spirit (denatured alcohol), two or three coats to the surface, allowing each coat to dry before applying the next. Do not apply undiluted shellac, this will simply blur any fine detail. When the surface is sealed and the shellac is hard, apply a coating of liquid wax evenly all over. The mould is now ready for filling. For polyester resin it is desirable to apply a thin coat of polyvinyl alcohol prepared solution over the wax.

Wax when used as a mould requires no parting agent. To remove a wax mould simply peel the wax off. It may be necessary to warm the wax first to soften it. If it is possible, immerse the whole thing, mould and cast, in hot water for as long as it is necessary to soften the wax to be able to peel it away. Alternatively, melt the wax and allow it to run off using a blow lamp or some such direct heat. Plaster, cements, concretes and synthetic resins can be satisfactorily cast from wax moulds.

Polyvinyl Alcohol solutions are manufactured for the plastics industry and are used on moulds of various kinds to make castings of polyester resins. They are best applied over a wax priming. The manufacturer of the particular resin used will recommend a suitable parting agent for use with their resin.

French Chalk or talc is a good parting agent. It is particularly useful for preparing the surface of a plaster piece mould from which a clay press cast is to be made. It forms an efficient release between clay and almost any surface. Because of this fact, be certain that no chalk gets between layers of clay during pressing or any other time. It will prevent clay adhering properly to clay, indicating its efficiency as a parting agent.

Graphite is applied to a plaster and grog mould surface to afford a good release and a good reproduction of the surface of a lead casting.

Rape Seed Oil comes from the colewort or coleseed plant and is particularly useful as a release agent for polyester resin from plaster of Paris. Coats of oil are applied to the dry mould over a period of days, till the mould is well impregnated. The resin can then be painted straight on to this. This material is probably the quickest, most efficient release agent for a plaster casting. The mould must be dry.

Tallow is a common release agent and is used on top of a sealed surface or a non-porous surface. It is particularly useful when piece moulding.

Oil is a simple and effective release agent. It can be used on a variety of moulds that require a water repellant surface. When hot, oil can be mixed with hot meat fat and this is still a favourite mixture of the jobbing plaster casters, being a very cheap and available fresh, whenever meat is cooked.

Expanded Polystyrene used as a mould requires no parting agent. This can be disposed of by heating with a blow lamp or by applying a solution of styrene. It is often used as a shuttering, allowing texture and modelling to be made *in situ*.

Polythene Sheet and other such plastic sheets are very useful materials to cast against. Cellophane or neoprene sheet is useful in the same way. Castings of almost any cold setting material can be made against them. Experiments in the studio will familiarise the sculptor with the possibilities these materials offer.

Wax polishes as liquids and creams are invaluable and can be used in a great variety of casting processes. I have already mentioned them in connection with treatment for moulds from which cement or synthetic resins are to be cast. Silicone waxes can be used in the same way and are, if anything, more efficient than ordinary wax polishes. The most difficult problem, however, with wax release agents, is removing them from the cast which is to be painted or is to receive some surface treatment. An effective de-greaser such as household ammonia has to be used.

Reinforcement of Casts

There are very few casting materials, other than metals, strong enough to reinforce a casting – an essential procedure. Most are brittle in their hard state and so require proper reinforcing. Jute scrim, glass fibre, Terylene fibre, carbon fibre, galvanised wire of various gauges, mild steel, high tensile aluminium rod, stainless steel and expanded metals are the most common materials used in sculpture to provide adequate strength to a casting. This strength is required at the points where the casting is likely to receive strain great enough to fracture the casting media. This is an important factor if a cast is to be placed on a public site, or where it is likely to undergo inordinate stress and strain, such as having small boys climbing over it, or students clambering on during a rag day celebration. These two events may seem unlikely, but such happenings have caught the unwary sculptor off guard, leaving him with a complicated repair problem.

It is advisable whenever possible to make castings which are hollow in construction and relatively light in weight. The tube is a stronger form than the solid rod and a hollow sphere has a proved strength; witness the egg. There was in the past a common practice amongst sculptors and casters of placing heavy mild steel reinforcement in the centre of a mass, making the mass solid by building up the material around the armature. This obviously can be a very quick method of filling, and was employed by jobbing casters when making castings in plaster, which were later to be reproduced in metal. These plaster casts often cracked, however, at such places as ankles, knees and wrists. Cracks appear, in fact, in those places where the forms are weakest and therefore vulnerable. It is true that with such reinforcement the forms only crack; they do not usually break. The cracks, however, have to be repaired and constantly made good. Cracking also occurs during journeys from the studio to the foundry and often whenever the casting is moved. Small figures do not suffer quite so much in this way, but a large work always receives some damage. The Rodin Museum in Paris shows some of the master plaster casts used by the foundry, and many of these carry the scars of cracking and repair. Rodin usually had piece moulds to hand, so that he could produce another plaster whenever he required. This led Rodin to develop his many varied techniques in assembling large sculptures. He often used one feature, e.g. a head, as on the *Burghers of Calais*, from which he piece-moulded and used a number of times in this one sculpture, making slight variations in the detail.

Cracking is due to the fact that outside surfaces of a form receive the greatest stress. If the form is strengthened only in the centre of the mass no benefit is given to the outer surfaces, stress in these areas will still be sufficient to cause cracking. In the case of concrete sculpture, or forms made in polyester resins, which are made to be placed in an open air environment, such cracking would prove to be disastrous. Weather penetrates the material and the sculpture would quickly deteriorate. This would not please a client, the sculptor would find himself constantly repairing it, and his reputation would also suffer.

I advise a hollow casting technqiue, reinforced on what I call the 'gaiter principle', using a great number of narrow gauge reinforcing rods which reinforce the actual cast material thickness, ensuring maximum strength to the outer surfaces, rather than placing one or two heavy gauge metals in the centre of the mass. Use, whenever possible, a suitable fibrous reinforcement, such as jute scrim in plaster of Paris, glass fibre in polyester resin and some concrete casting. Terylene fibre can be used in the same way as glass fibre. Incorporate these fibres into the cast thickness together with metal reinforcement. Do not be tempted to make a large form solid or part solid. Small forms that are unlikely to receive undue stress, however, can be made solid. I personally make most castings hollow as, in the long run, this is a much more satisfactory job. In the case of concrete castings, hollow forms, made whenever possible with an open end, withstand the weather most satisfactorily. Moisture will soak

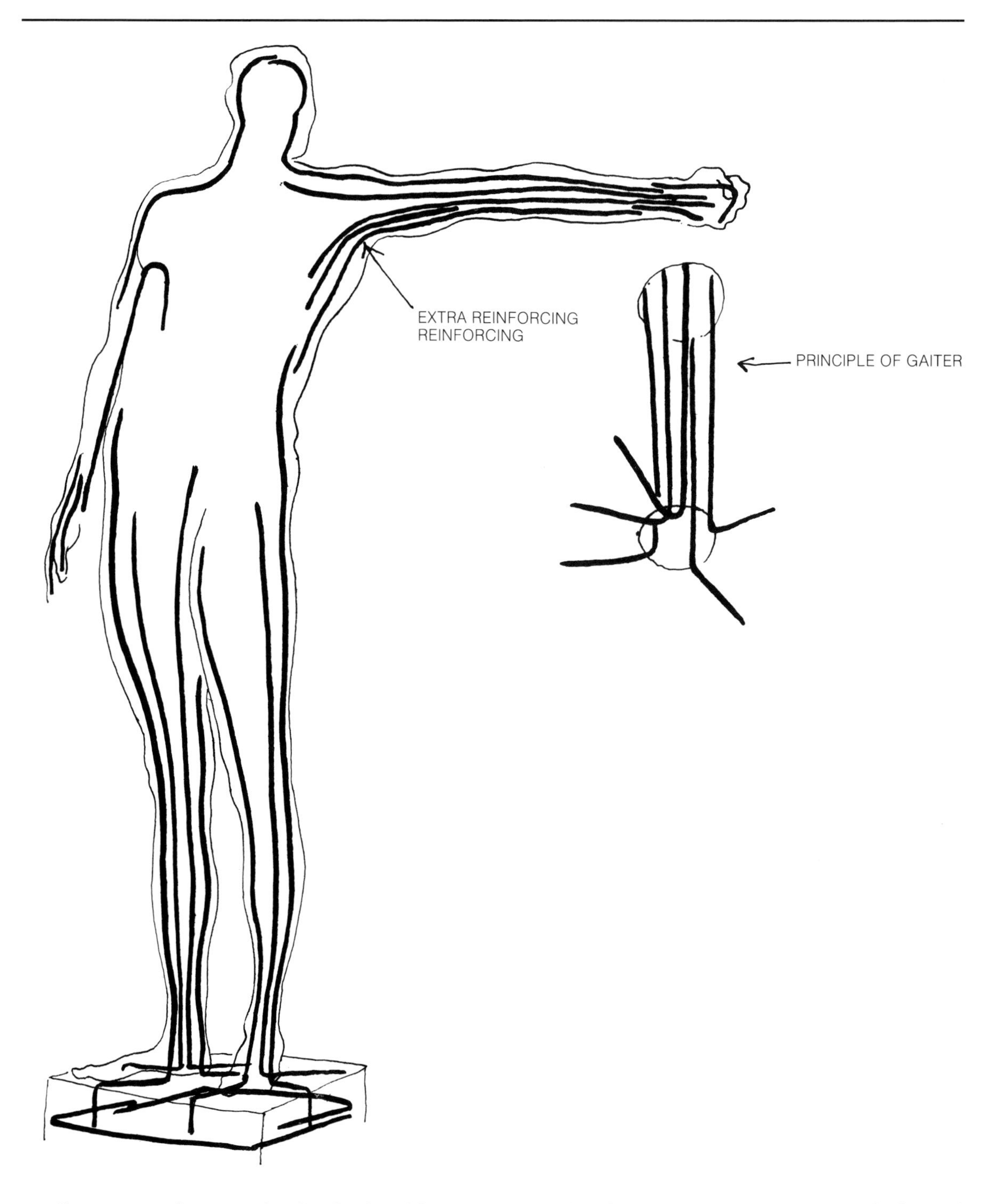

54 Illustration showing the kind of mild steel reinforcement required in casting. The irons are placed around the casting in a form of gaiter; this affords the maximum support to the surfaces that will take most of the strain placed upon the cast. Irons placed at the centre of forms do very little good and will not stop the surfaces of the cast from cracking, especially around such slender forms as the ankles.

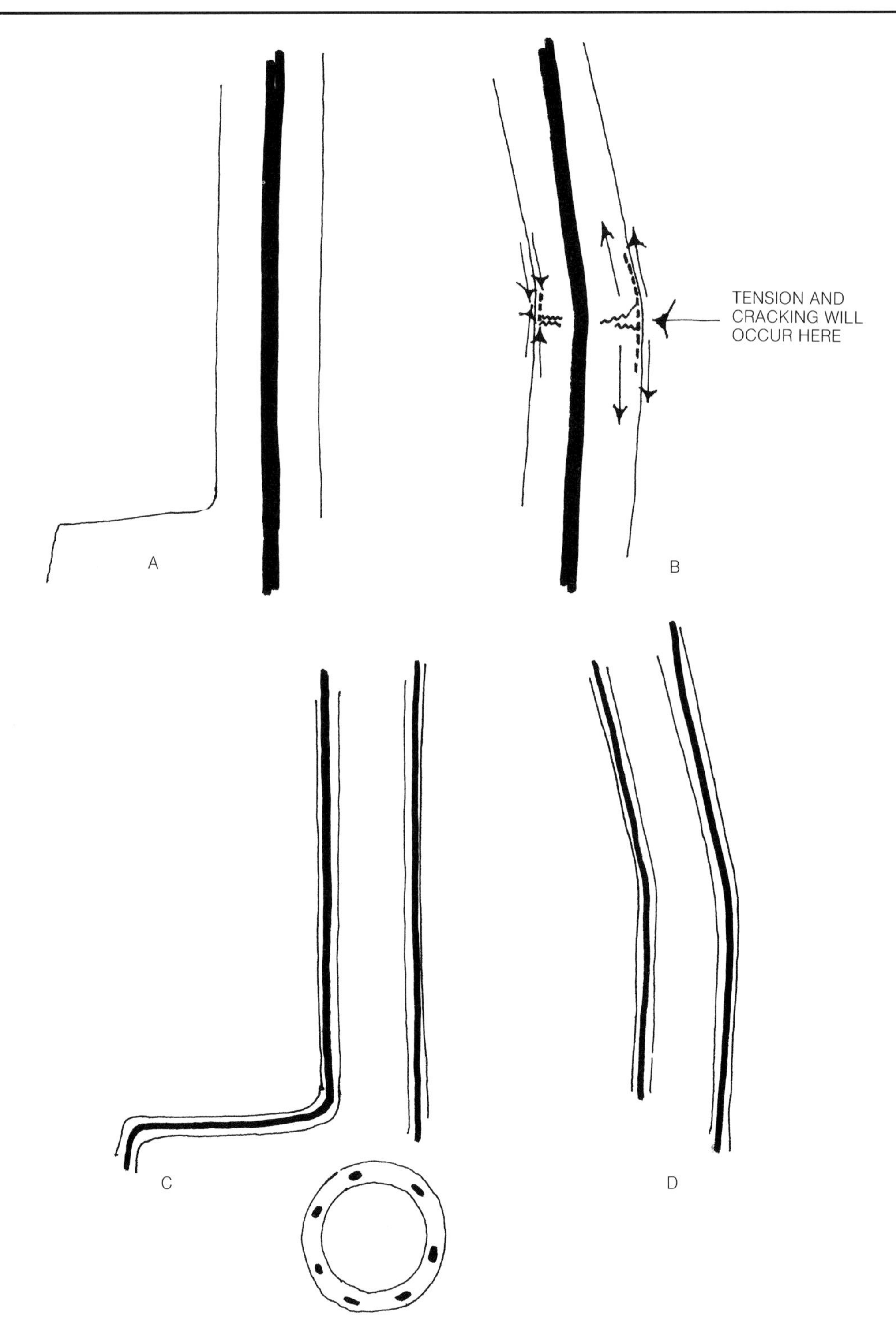

55 Iron reinforcement. A and B: a single reinforcement showing the vulnerability of the surfaces. C and D: reinforcement placed within the cast thickness of hollow casting, giving greater strength.

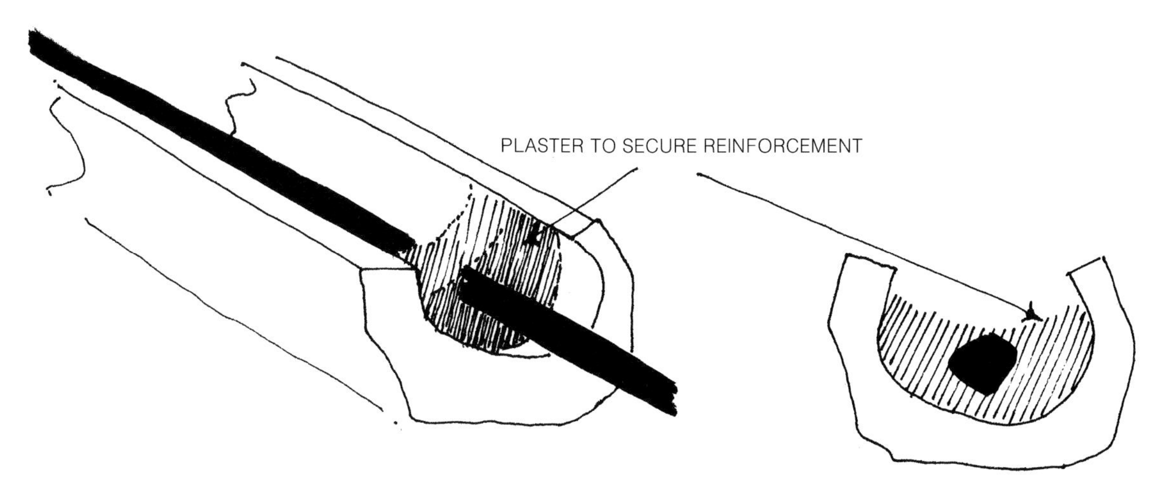

56 Mild steel reinforcing secured simply in the mould by small applications of plaster. Care must be taken not to block the opening through which the filling is poured.

through and out of hollow forms fairly quickly. Water, if retained in the material, can damage the casting if it freezes, causing cracking and breakage.

Get to know your materials; weigh up the pros and cons before casting a specific work; plan the strongest possible job. It is worth taking more time to place sensible reinforcing than to suffer setbacks later. A study of organic structure will help to gain insight into various forms of strength. A study should also be made of reinforcement used in the building industry, since this will be extremely useful. Strength of a tensile character, rather than compressive nature, is the principal concern of the sculptor today, and such tensile strength is embodied in an enormous field of research and development in all branches of industry.

Filling the Mould

Plaster of Paris

Prepare a number of moulds, of the various basic forms, to be filled with plaster of Paris. For this material the mould must have a coloured warning layer and have a sufficiently wide opening through which to pour a liquid filling. The process for making a solid plaster of Paris casting is as follows:

1. Prepare the mould, using soft soap or liquid detergent as a release agent.
2. Soak the mould. The mould is sufficiently wet when water remains on the surface of the plaster after being removed from the tank. If it is possible to immerse the mould, it will be wet enough when the air bubbles stop rising from it.
3. At this point any reinforcement necessary must be made. This should be placed in the main mould and held in position with little mounds of plaster, just enough to secure the metal but not obstruct the pouring. Mild steel, if used to reinforce plaster, must be painted with a preparation to prevent it rusting, such as galvanising paint, shellac, wax and amber resin or Brunswick black paint.
4. Close the mould by placing the cap or caps to the main mould. Secure these by using scrim and plaster. Seal all round the seam with plaster to prevent any leaks from these points. The mould can be secured with string or cord and the caps tied and held in position. The seams can be sealed with

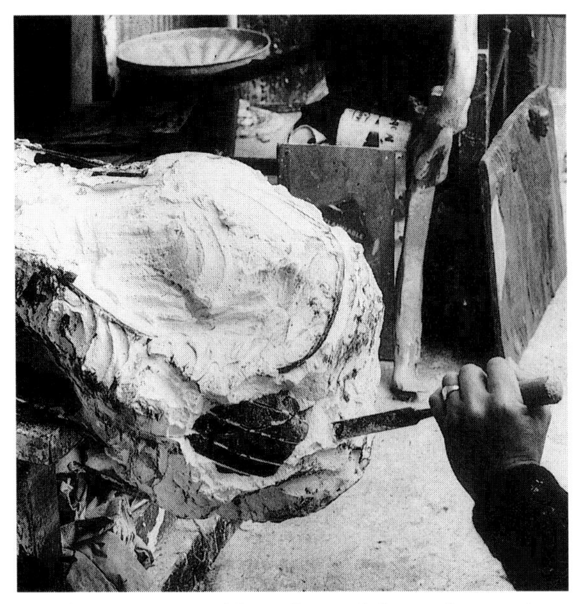

57 The waste mould is chipped from the casting, commencing at the highest point.

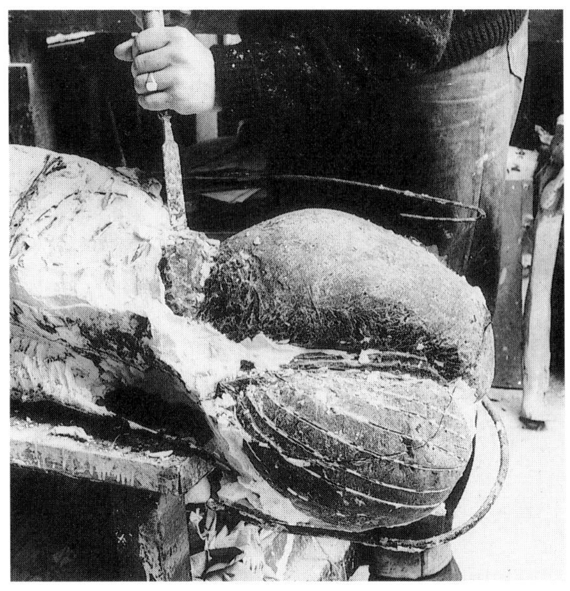

58 Gradually the mould is chipped off to reveal the cast.

clay. I prefer the former technique.

5 When the mould is closed and is secure and the plaster seal hardened, it is ready to be filled. If the mould is large and time elapses during the reinforcing, closing and securing processes, then soak the mould again. Immerse the mould if possible, but if not, then spray it with water.
6 Place the prepared mould with the opening presented through which to pour the filling. Prop it up to prevent it tipping, and, if possible stand it in, or over, a bucket or bowl. This will prove useful should the mould spring a leak.
7 Mix sufficient plaster of Paris to fill the mould.
8 With an assistant standing by armed with a lump of soft clay, ready to plug any leak, pour plaster into the mould to fill it. Pour gently and steadily into one opening. Vibrate the mould gently by tapping with the palm of the hand. This will help to release any air bubbles. When the plaster stiffens, draw a straight edge across the opening to level the filling.
9 When the filling has set and hardened, turn the whole thing and stand the mould the right way up. This allows moisture to drain towards the bottom to make chipping out much easier. Leave it to stand for at least an hour, longer if possible. After this the cast is ready to be 'chipped out'.
10 Chip out with blunted joiner's chisel and wooden mallet. Commence at the top of the mould, holding the chisel at right angles to the surface of the cast. Chip away the waste; the coloured warning coat will now indicate how near to the cast surface the chisel is. When chipping out, always reveal the high points of the cast, and gradually and carefully work back from these. Always direct the chisel towards the bulk of the cast, never across a form, especially a fine form. Resist the temptation to prise the mould off. With plaster of Paris this will result simply in loss of detail. The chipping out is complete when every vestige of mould has been removed.
11 The casting now has to be cleaned up, any seam flash removed and any defect in the

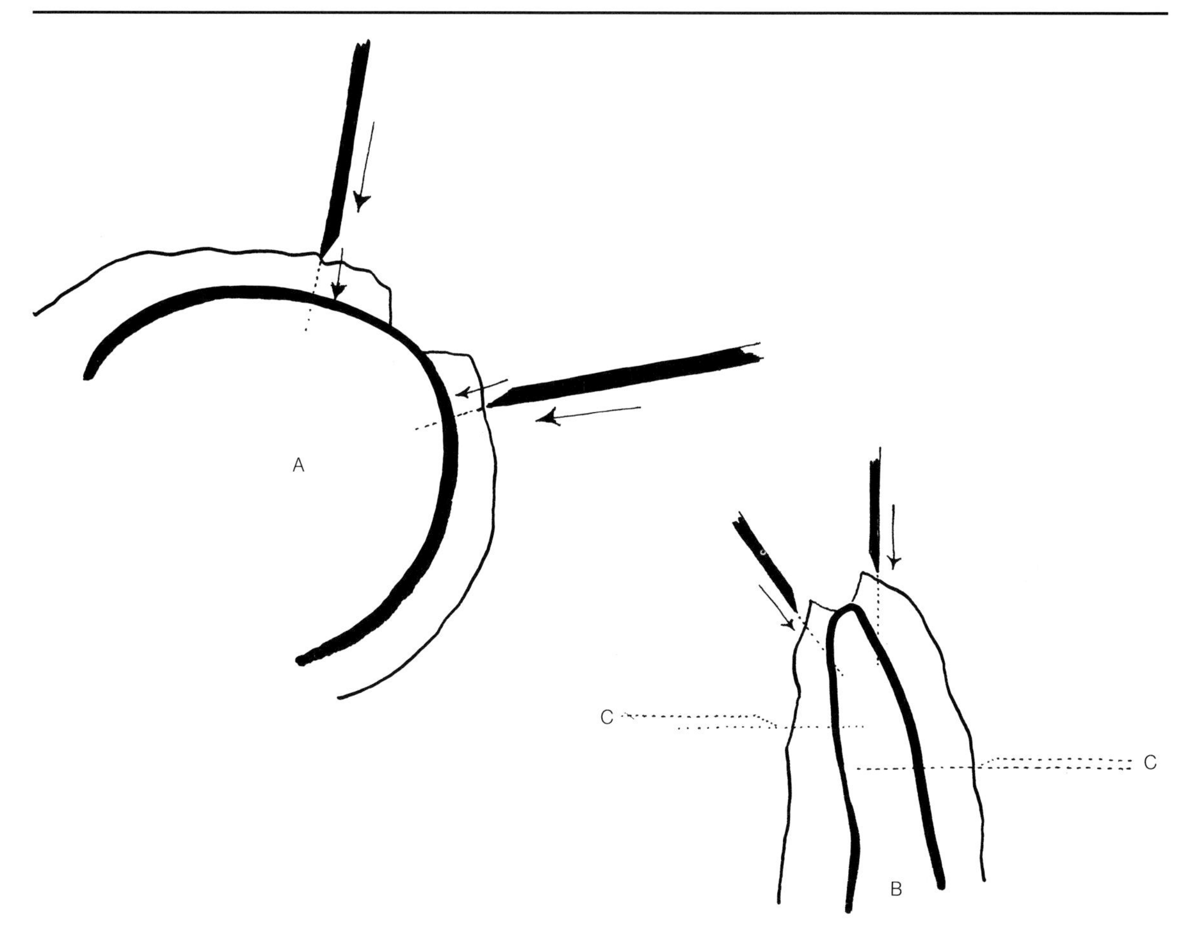

59 Chipping out. A: the direction of the chisel to cast surface when chipping out. B: the direction of the chisel chipping into the mass of slender form. *Never* direct the chisel as indicated at C.

cast repaired. Allow the cleaned and repaired cast to dry and harden.

Any of the basic forms can be cast in this way to produce solid castings. Points to remember are soaping adequately and soaking. If either of these two processes are missed, no matter how well other things are done, a satisfactory casting cannot be made. Insufficient soaping results in adhesion between mould and cast, and unless the mould is soaked well it will result in a weak cast which will have an aerated surface, and crumble.

Large solid castings are often undesirable for reasons of size, weight or expense. Hollow castings are therefore made. The process for making hollow castings from a suitable mould is as follows:

1 Prepare the mould surface with a solution of soft soap or liquid detergent. The moulds need to be designed to allow access to the whole mould surface.
2 Soak the mould, and prepare any reinforcement. See pages 53–56.
3 Fill the main mould and caps by applying evenly to the surface of the prepared and soaked mould a layer of plaster of Paris, approximately $\frac{1}{4}$in (6mm) thick. Be careful not to obscure seam edges; it is best to

sponge these edges to keep plaster off them. Key the back surface of this first layer, and when it is hardened apply a second layer with jute scrim reinforcement. This is best laid by cutting the scrim into lengths, dipping them into mixed plaster and placing them carefully in position. The thickness of the two layers need be no more than $\frac{1}{2}$in (13mm). Any mild steel or other reinforcement needed, should be placed and fixed at this time with scrim and plaster.

4 Trim the seam edges and make sure the caps fit to the main mould. Make any adjustments necessary to ensure a good fit.

5 The method of fixing the caps to the main mould, thence to complete the filling, can be one of, or a combination of three.

1 Pouring
2 Squeezing
3 Packing

1 Pouring requires the caps to be fixed in position, with scrim and plaster, and the seams sealed and plugged. A mix of plaster is then poured into the mould; poured out and in again; it is then rolled around allowing the fresh mix of plaster to fill the seam all round. The rolling action is continued until the plaster sets. This fills the seam and at the same time secures the cast pieces. Of course this method depends very much on the size and manageability of the mould and whether or not it has an opening through which to pour the plaster in and out.

2 Squeezing is the method used when the mould has no opening or is too bulky to pour and roll. The filling is placed close up to the seam edges. Make sure the caps fit well to the main mould with the filling. Make any adjustments necessary. Now at the seam edge plaster is added to caps and main mould, and the caps are squeezed onto the main mould. Surplus plaster is forced out all round the seam and at the same time forced into the cast. Caps can be gently tapped home with a wooden mallet to ensure the closest possible fit. Do not try to cope with more than one cap at a time. Any plaster remaining from one cap may obstruct the proper placing of the next and must be trimmed off.

3 Packing is the method used when the mould is large and bulky and has an opening, such as is offered by pieces of mould along a large cylindrical form. Access to the seam on the inside of the mould is therefore possible before placing the next cap. The filling is placed in up to the seams of the caps in the main mould; it is then trimmed at the seams, and the caps are placed one at a time. The space at the seam, caused by the cap and main mould being placed together, is then packed with plaster from the inside, scrim being added to strengthen the bond. In this way the casting is completed cap by cap, each seam being cast and strengthened with care. The process is the same for each cap. If the form is closed, the final cap must be squeezed; if the form is open, it can be packed.

6 When the filling is complete allow the cast to stand the right way up to harden evenly and satisfactorily. Then chip off the waste mould.

Any basic form can be cast in this way to produce hollow castings, so select the method of filling most suited to the forms.

Points to remember:

1 The necessity for even coverage of the mould surface with the first layer of plaster, including high points (low points and hollows on the positive form). This must be followed up with an even covering of scrim and plaster and any suitable reinforcement.

2 When pouring and rolling, ensure that the plaster is rolled round the seam. If it is possible, include some scrim across the seams.

3 When packing, make sure the seam is accessible on the inside and very thorough.

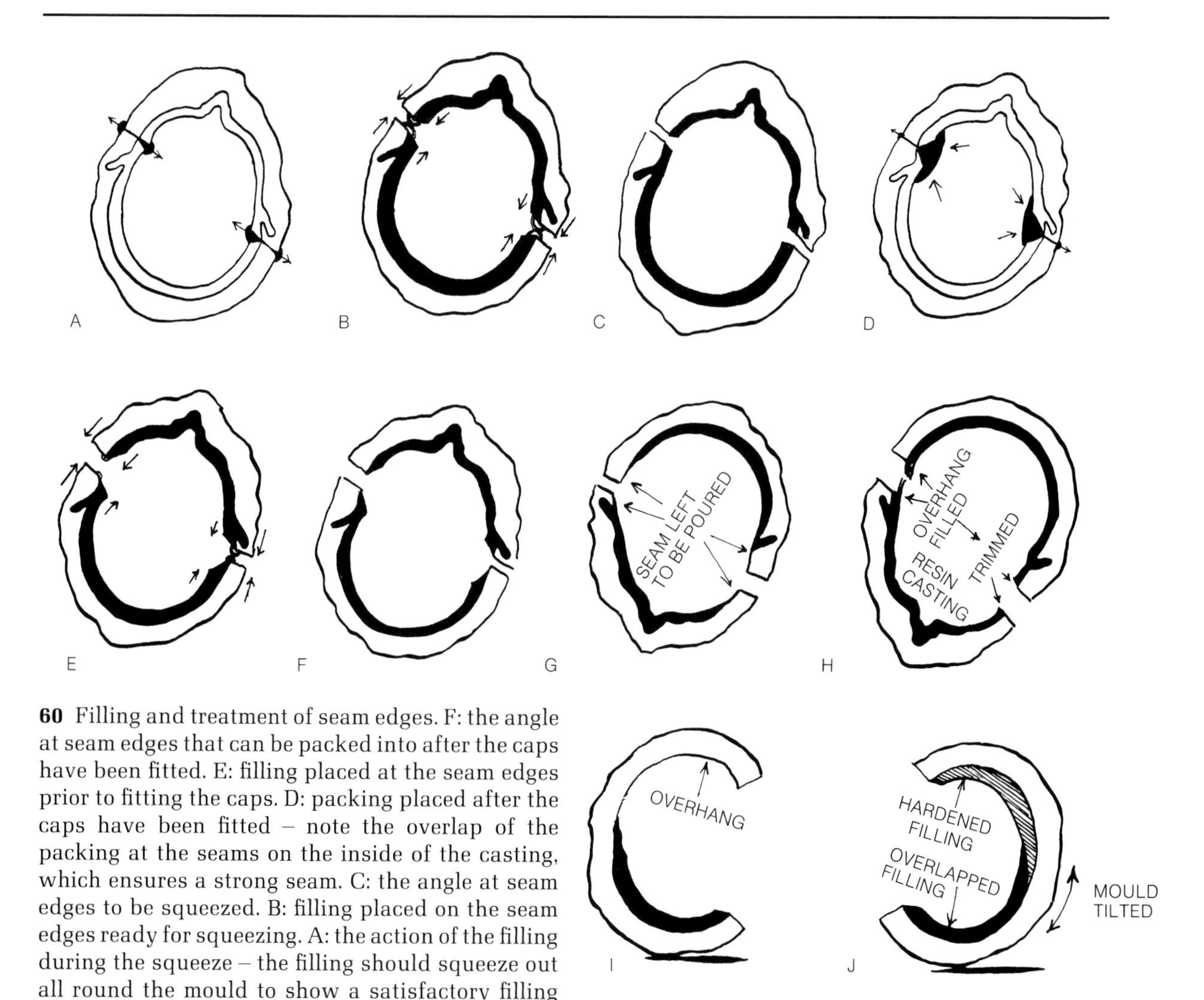

60 Filling and treatment of seam edges. F: the angle at seam edges that can be packed into after the caps have been fitted. E: filling placed at the seam edges prior to fitting the caps. D: packing placed after the caps have been fitted – note the overlap of the packing at the seams on the inside of the casting, which ensures a strong seam. C: the angle at seam edges to be squeezed. B: filling placed on the seam edges ready for squeezing. A: the action of the filling during the squeeze – the filling should squeeze out all round the mould to show a satisfactory filling and must not be too liquid or it will run down into the casting leaving nothing at the join. G: seam edges avoided during packing to keep them clean and ensure a good fit – seams will be filled later by pouring and rolling the material around the seam. H and J: the treatment of overhanging surfaces. I: the filling and trimming at seam edges, for resin and fibreglass – this is done with a sharp knife or with fine hack saws.

4 Squeezing requires an even distribution of the build up around the two seams and an even squeeze.
5 Do not disturb caps as the plaster hardens.

Practice of course can make masters of most of us, and it is useful to experiment with all the methods described. Complex forms may involve a combination of these techniques, so try to master them all before taking on the casting of a sculpture which is important to you.

Piece Moulds are filled in the same way. The preparation of the mould, however, is quite different, and must be carried out very thoroughly. This is best done by sealing with shellac. The surface must be dry. Apply the shellac as a fairly dilute solution with methylated spirit (denatured alcohol). It is better to seal a surface by applying successive coats of dilute solution, than by putting on an undiluted shellac, which will only collect in the hollows and spoil the fidelity of reproduction. Wax or tallow can then be painted on to the then sealed surface to make an efficient release. Fill the mould by whatever method is most suitable to the form, size and character, to produce a satisfactory casting. When this filling has hardened, remove carefully and in the correct order the pieces of mould. Clean each piece as it is removed and place it carefully into the mould case which must also be free from any loose particles of plaster, or dirt of any kind. The slightest speck caught between pieces, or between the case and piece, will spoil the correct register of the seams and cause an inaccurate casting which will require a great deal of work to put right.

Cement and Concrete

It is possible to fill an empty mould with a cement or concrete mix. Having done this vibrate the mould to release any air bubbles from the casting, allow it to set and cure, then chip away or remove the mould. Reinforcement can be simply pushed into the wet filling, or fixed prior to pouring the filling. This is a simple method of casting, crude and limited in the degree of refinement possible to achieve. Industry has developed this simple method to a very sophisticated degree, and it is applied particularly to the building industry. The principle is the same as described above and it is obviously similar to simple filling of a mould with any liquid casting substance. Terms in the building industry, however, differ. The mould is a shuttering, and mechanical vibrators are used to evacuate air bubbles to produce a fine surface. The concrete used is a mixture very different in characer to that used by sculptors. Most buildings today incorporate structures built in this direct way. The student should investigate some of these structures and techniques to discover the possibilities and qualities inherent in them.

For casting with concretes to give the greatest possible fidelity of reproduction to the original, techniques are necessarily more refined. Developments and adaptations of the studio methods used for plaster of Paris moulding and casting have taken place, producing strong hollow castings. According to the nature of the sculpture an amalgam of industrial methods and studio techniques is possible.

Some years ago I became interested, along with many other sculptors, in the possibilities of cements and concretes for sculpture. Costs of casting in bronze had reached, and remain still, at an exorbitant level. It was the concern of young sculptors to find a material that would enable them to model a sculpture freely, then reproduce it in a substance that would permit its use in an architectural site, thereby enabling sculpture to be used in public sites at reasonable cost. Of the materials, cements and concrete are possibly amongst the most widely used. Buildings are constructed largely from concrete, and there is no reason why sculpture designed for architectural sites should not be of concrete too. But the castings first produced were often heavy and clumsy, making transportation very difficult, apart from causing complications in the studio. The techniques used originally were similar to industrial methods, but carried out with great care. The possibilities of the material were explored, however, and artists could show and sell their work in galleries, certain in the knowledge that the material used would have a certain long life. Commissions were given and concrete gradually crept into the repertoire of the skills of sculpture. But the big disadvantage to most was still the inordinate weight of the final work. Lugging it around the studio and transporting it to the site or gallery, all took time, equipment and money. What was needed in addition to the castability of concrete, was a technique which would enable a casting to be made so that it would be strong and durable, and at the same time light in weight.

Together with one or two colleagues, principally Sydney Harpley, I set about developing such a method, during my studies at the Royal College of Art. A technique was developed finally that gave the quality of strength and durability, and the satisfaction of light weight. It gave in addition a greater range of surface texture and colour, and additional tensile strength. This method, together with the traditional concrete casting technique, is now very widely used in Britain, and as a result many sculptors have been able to sell their work. Consequently more sculpture has been used by architects and local authorities. I was pleased to introduce this casting material and technique to students at the Universities of Michigan and Eastern Michigan, enabling students to gain valuable casting experience with the bonus of making a sculpture in a durable material. I will describe the methods indicated above as well as the traditional technique. It is essential to my mind to have an adequate knowledge of both.

The mould for producing a concrete casting must be designed to allow access to the total mould surface. The technique is similar to that employed to make a hollow plaster of Paris casting. The moulds when complete must be prepared according to the work. For cements and concretes the simplest preparation is the clay film left on the mould surface, when the clay original is removed. Alternatives are shellac and wax (see release agents page 52).

I recommend a student to make as many moulds as possible and use them to make castings, from which valuable experience can be gained.

When the mould is prepared, mix the cement and aggregate. The mixture is a very important factor in the process of casting with this material.

Cement This is the binding agent for all mixes. It can be used neat (without an aggregate) when using the techniques I will describe adding glass fibre. There are two basic types of cement:

1 *Calcium aluminium silicate*, which is used to make the Portland cement types most common in the building industry.
2 *Alumina*, which is the raw material from which aluminous cements are made (Ciment Fondu in Britain and Europe, Lumnite in America).

Aggregates These are the materials mixed with cement, which give the final set, hard material its particular strength. There is a saying that 'a mix is only as strong as the aggregate'.

Mortar This is a mixture of sand and cement, used in building to lay bricks and make cement rendering.

Concrete In the building and construction industry, this name is given to the substance resulting from a mixture of cement, a small aggregate such as sand, plus a large aggregate with a gradation in size of particles, from coarse sand up to large stone. It is used, for example, for floors and columns where great compressive strength is required.

Sculptors tend to lump castings of these various mixtures under the generic heading of 'concrete'. Mortar and concrete are terms more applicable to industry, together with reinforced concrete and pre-stressed concrete. The latter two materials are not often employed by sculptors; their title, however, may need some explanation. Reinforced concrete results when the mixture has been poured into a mould (shuttering) and steel reinforcement included in the casting to give tensile strength. Prestressed concrete is concrete poured into a mould (shuttering) in which steel reinforcement is already placed and retained under tension. The resulting cast form gains greater

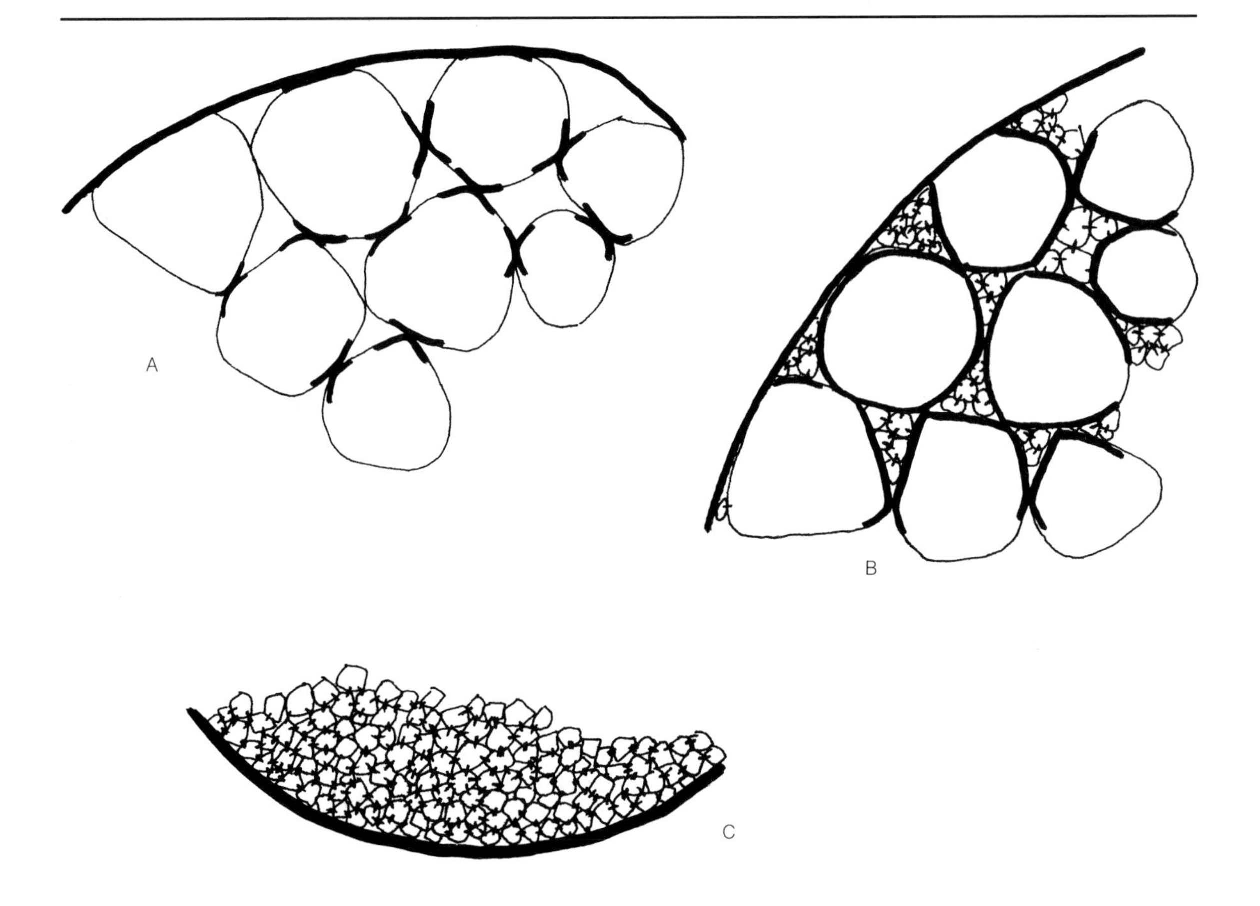

61 Aggregate surface area. A: the smaller surface area of a large aggregate. B: the variable surface area of a mixed aggregate. C: the greater surface area of a fine aggregate.

tensile strength in this way and enables beautiful, fine forms to be used in buildings.

Cements have a setting speed, a hardening speed and a speed of cure. These are important factors in the selection of material for a particular work. Generally Portland type cements are cheaper than aluminous cements. There are proprietary additives for all types of cements, for colouring, to prevent dusting, and to affect the setting rate. Advice regarding the use of these additives should be sought from the dealer supplying the material. He should be able to advise on the mixture and advisability of mixtures according to the particular cement and the conditions in which it is to be used. Generally the relevant setting, hardening and curing times, for cements without additives, are as follows:

	Setting	*Hardening*	*Curing*
Portland	10 hours	3 days	21 days
Portland rapid hardening	10 hours	24 hours	7 days
Aluminous	3 hours	7 hours	24 hours

The additives mentioned include a variety of colouring compounds; these can be added to all kinds of cements to change and control the colour. Advice should also be sought from the dealer regarding these additives. Portland cements are a pale grey colour, which may vary according to the colour of the aggregate. Aluminous cements naturally are a very dark grey, almost black; this colour too may vary with the aggregate. Interesting ranges of colour and texture can be achieved by using various aggregate mixtures, which may affect the general colour, or be exposed to reveal the particular aggregate pieces. This is a technique used extensively today to make interesting cladding, giving colour and texture to building. The aggregate can be exposed by grinding and polishing the final set, by sand blasting the final set or by removing the cement fines from the surface before the final set and cure takes place.

Sizes of aggregate vary from very fine powder almost as fine as the cement itself, up to a $\frac{1}{4}$in (6mm) or more. This size is known as the *sieve size* of the aggregate, being related to the mesh size of the sieve used to separate the various particles of aggregate. Coarse aggregate affords the greatest compressive strength to a mix, hence the reason for the large stone aggregate used in building foundations for example.

Being the binder, cement when added to the aggregate to make the mix is used in ratio to the surface area of the aggregate. The smaller the aggregate size the greater the surface area. The greater the aggregate size the smaller the surface area. The shape of the aggregate pieces should be as granular as possible, made up of facets which can be stuck together by means of the binder. The surface area therefore is measured by the facets or faces at which the aggregate particles join. Proportions are important to the workability, strength, contraction and general quality of the substance in its final set state.

Coarser aggregates, because of their small surface area, are mixed in the ratios of 1 part cement to 6 parts of large aggregate (2 parts gravel, 4 parts sand); this is a concrete. A mortar mix using only sand as an aggregate is 1 part cement to 3 parts sand; 1 to 3 because the surface area of the aggregate is greater. Finer aggregates such as pumice powder, or brick dust, or very fine mesh washed sand, are mixed in equal proportions, because the surface area of the aggregate is larger still. It is unwise to increase the proportion of aggregate when its sieve size is almost the same as the size of the cement. The perfect aggregate contains particles evenly graded between the largest and the smallest, and is clean and dry. Always use a consistent measure throughout a job to be certain of mixes.

The water added to make the mix should be clean. This should be added proportionate to the amount of cement in the mixture. The more cement used the more water it is possible to add, retaining the strength and character of the mix. Water added in too great a quantity to a mixture, with a small proportion of cement, will simply filter the cement down through the aggregate. The wet mix should never be liquid, however. If, when using sand or larger aggregate, it is necessary to make a wetter mix than usual, then add more cement. The actual amount of water added to make a mix workable varies according to the nature of the work.

The coarser aggregates naturally are rugged in appearance and should be selected with this character in mind. Sand as an aggregate can be used to achieve both a fairly fine casting or a more rugged open character. To achieve fine fidelity surfaces, together with the compressive strength of large aggregates, a mixture of finer material (neat cement for instance) is applied to the surface of the mould first. Fine aggregates afford reproduction of the finest detail; finger prints can be easily reproduced. To gain greater strength from the fine aggregate mixes, it is wise to add, in layers or laminations, glass fibre or Terylene fibre. This will serve to give the casting greater tensile strength, and produce a very durable, lightweight final product. Of course these various mixtures can be combined and should be selected according to colour, appearance, compressive or tensile strength. When selecting and purchasing an aggregate, take care that it does not harbour

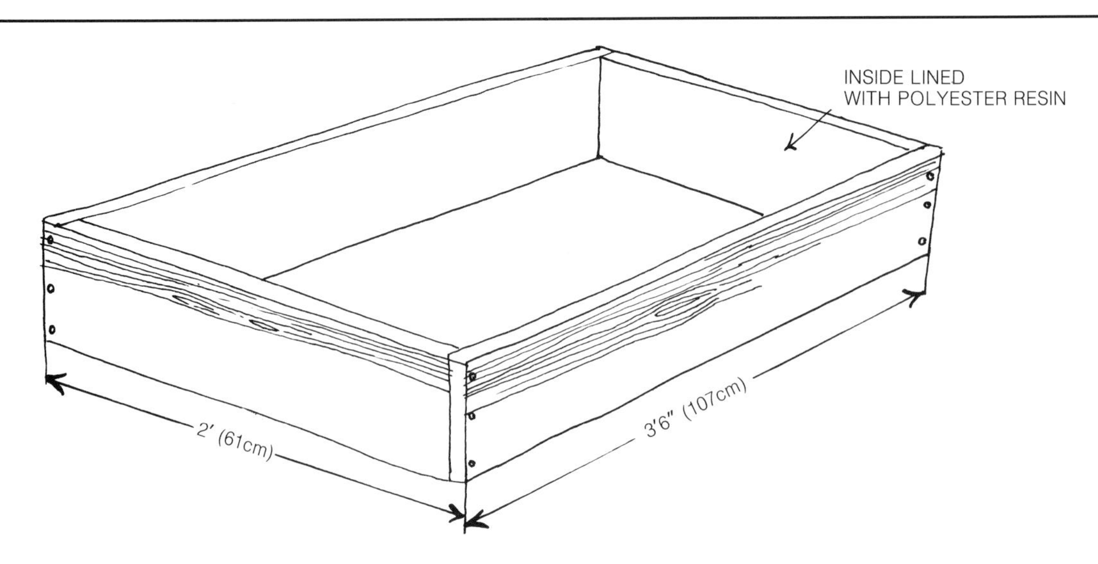

62 Concrete mixing box, made from wood with the inside surfaces painted with polyester resin to make it waterproof. This box is ideal for mixing both wet and dry concrete.

any deleterious material that will contaminate the cement and prevent it setting. This is particularly important when dealing with aluminous cements. Keep all the ingredients of a mix, including the water, as clean as possible.

Mixing, or making the mix to be used, is a most important factor in the process of casting. First of all, estimate the total amount of material required, to complete the job in hand. Measure the proportions accurately, using a consistent measure, a bowl or bucket or some such vessel will do. *Dry mix* these materials thoroughly till the colour of the dry mix is even throughout. Make sure there are no pockets of unmixed ingredients remaining. The *wet mix* is now made to be used in the mould. From the dry mix take enough material to deal with a particular area of the mould, or to complete a particular session of work. Try to estimate how much can be done at any one time. This varies of course according to the speed at which individuals work. It is not wise to wet mix the whole dry mixture; for a large mould this will only result in waste. For a small mould it may be safe to mix the total amount; however, this depends on the speed at which the sculptor works. It is possible to complete a small casting in a single session.

To make the *wet mix* add water gradually, mixing thoroughly till the consistency of the mix is as required. This consistency varies according to the aggregate. With a large aggregate, sand and other coarse materials, giving a small surface area, the mix should be fairly dry in character. Pat the mix and it should become moist on the surface. When squeezed in the hand this mix should remain firm in the forced shape. If a wetter mix is required, add more cement to compensate for the additional water. Smaller aggregates, offering greater surface areas, need more water to make a satisfactory mix. A consistency equal to a good thick cream is most satisfactory. For the cement and fine aggregate mix, this can be made similar in consistency to clay. This is useful when filling a mould, to ensure an even coverage of the mould surface.

I must stress again the importance of mixing both dry and wet mixes evenly. An even colour, without patches of unmixed aggregate, is the aim. Any unmixed patches will usually appear on the surface of the cast, and ruin the quality of the work. Both mixes can be satisfactorily made if a suitable vessel for mixing is used. I find the most useful item for this purpose is a baby's bath. These usually are a very manageable size and without corners. I use one to make the dry mix and another, or a plastic bowl, to make the wet mix. Failing this, a mixing box can be made and the inside painted with polyester resin and glass fibre, to make a satisfactory, resistant surface.

The following is a list of mixes I have found most useful. There are many more, however, and experiments should be carried out to test a new mix, or a new batch of materials.

- 1 part cement to 3 parts sharp sand.
- 1 part cement to 3 parts silver sand.
- 1 part cement to 1 part marble dust plus 3 parts marble chips from $\frac{1}{8}$in (3mm).
- 1 part cement to 1 part marble dust plus 3 parts granite chips from $\frac{1}{8}$in (3mm).
- 1 part cement to 3 parts grog (ground ceramic).
- 1 part cement to 3 parts ground stone (any good quality stone).
- 1 part cement to 1 part pumice powder plus 2 parts plastic granules (useful for exposed aggregate work).
- 2 parts cement to 2 parts marble dust plus 1 part pulverised fuel ash.
- 1 part cement to 1 part pumice powder.
- 1 part cement to 1 part brick dust.
- 1 part cement to 1 part mixed brick dust and marble dust.

Experiments should also be made with combinations of various mixes.

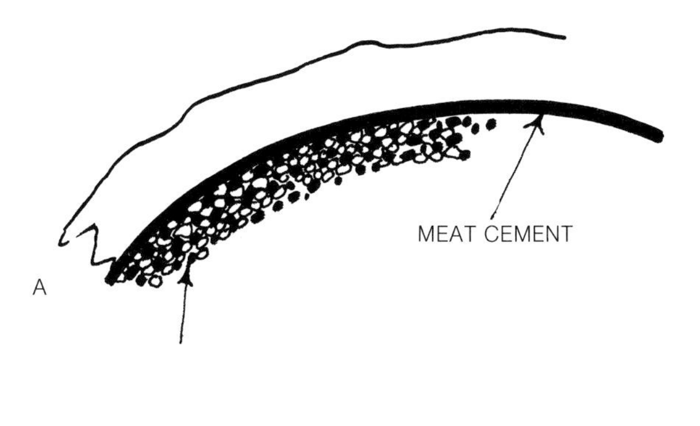

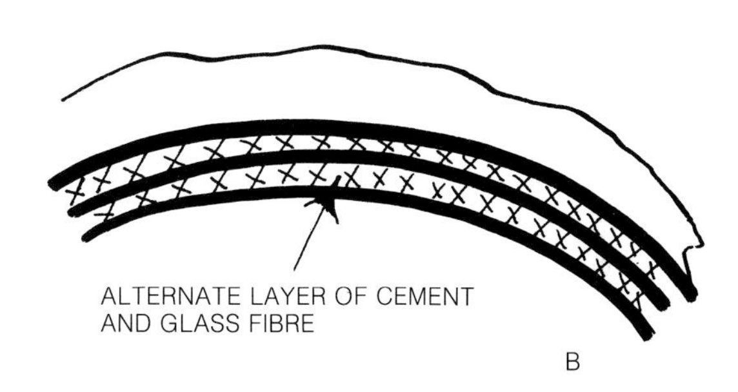

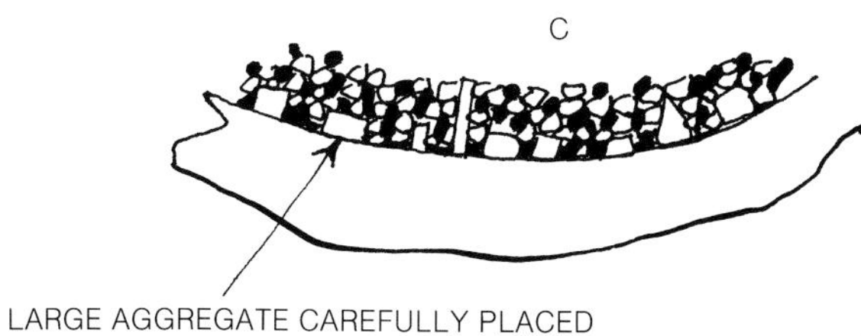

63 Various fillers. A: cement and sand with fine quality surface. B: cement and glass fibre. C: large aggregate to be exposed. D: resin and glass fibre lay up.

The process of filling is as follows:-

1 Prepare the mould with the proper release agent. Then soak the mould.
2 Prepare the concrete mix. Dry mix then wet mix.
3 Drain excess moisture from the mould. Ensure that the mould is wet all over.
4 Apply the concrete mix to the mould surface. This process of course varies according to the nature of the concrete:

   a Large aggregate mixes, requiring no surface detail, can be placed directly to the mould surface. This must be well consolidated to make a strong casting and this is achieved by tamping the material with the hand or with a wooden mallet. Build up the required thickness ½in to 1in (13mm to 26mm) over the mould surface in this way, tamping the material to consolidate it, till it seems to ring. Overhanging surfaces can be filled separately, by rolling the mould, when the packing so far applied, has

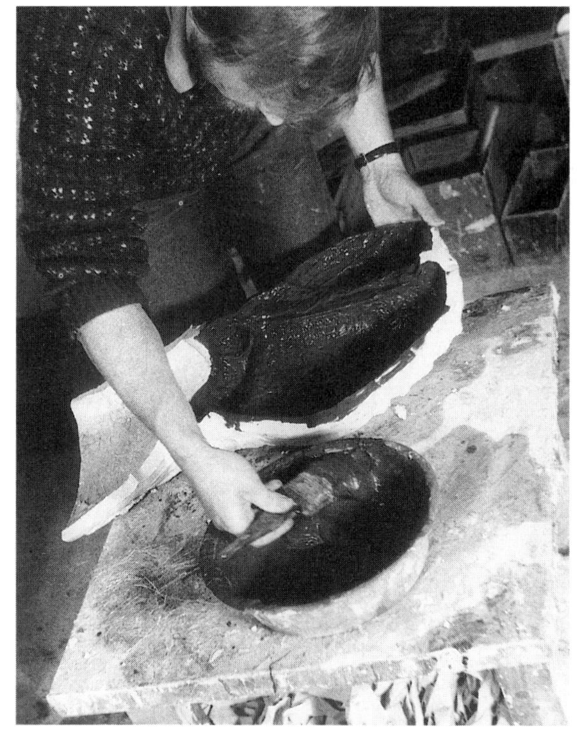

64 Applying the first layer of cement and fine aggregate to the mould surface.

65 Placing the glass fibre.

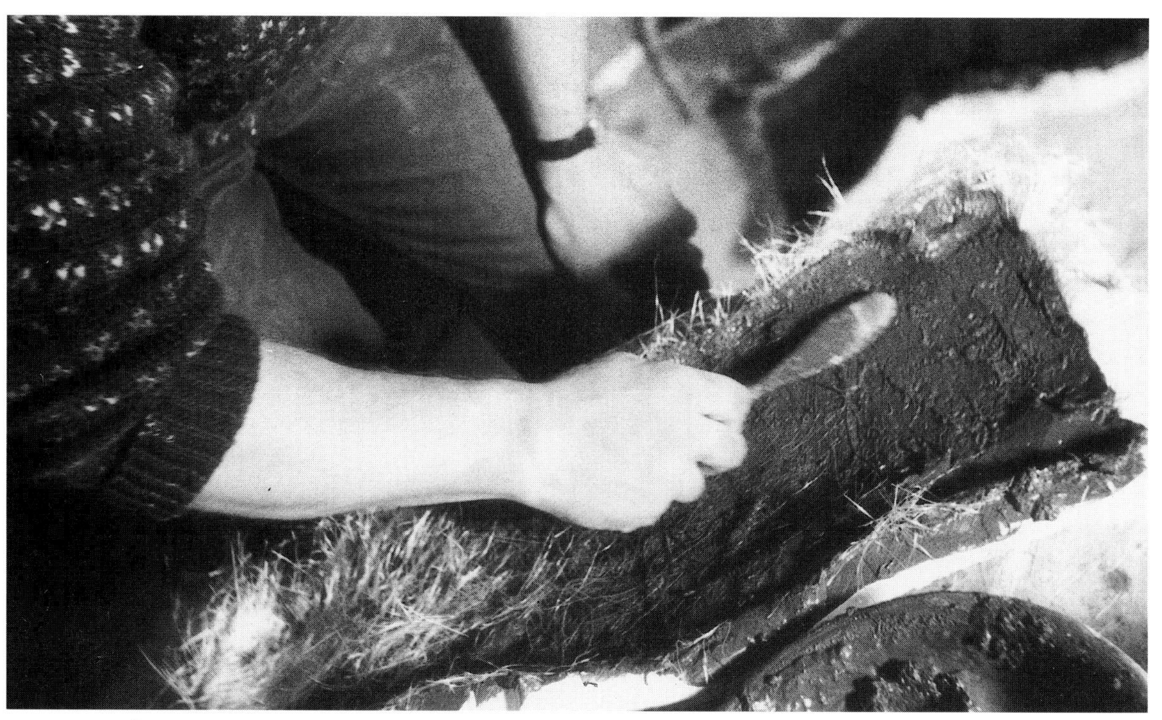

66 Applying the second layer of cement and fine aggregate.

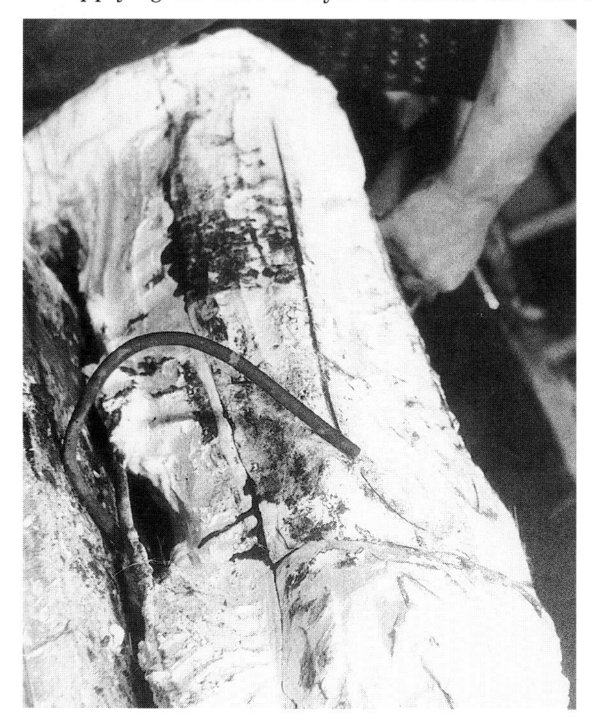

67 Caps squeezed to the main mould and held with a mild steel clamp. The seam was subsequently filled and reinforced from the open end of the mould.

hardened.

b Large aggregate mixes used for compressive strength or which are to be exposed, but which require fidelity of reproduction on the surface of the cast. It is necessary to apply to the mould surface a mix that will give a faithful surface reproduction. This may be either a first application of neat cement, or cement mixed with a fine aggregate. This application needs to be $\frac{1}{4}$in (6mm) thick and as even as possible. It can be less substantial if the casting is to be rubbed down to expose the aggregate. The drier mix, using the large aggregate, is built up on top of this first layer whilst it is still wet. The cast thickness is built up in this way with the coarse mix, which must be well tamped down to consolidate the cast build up.

c Fine aggregates or neat cements should be mixed to a thick cream consistency, as advised. A first application of this is made evenly over the mould surface, approximately $\frac{1}{4}$in (6mm) thick. On top of this a layer of glass fibre (chopped strand mat) is then placed. It will be possible to divide the mat, as delivered, to make a thinner mat. This fibre is worked into the surface layer, with a stiff brush, using a stippling action. The fibres must be impregnated with concrete. Then apply another layer of concrete or cement, then another layer of fibre. Continue this laminating process until the required cast thickness is made, the final surface being cement, made smooth with a trowel or spatula, to create a double surface tension of the material itself.

5 Any reinforcement, such as mild steel, should be included in the process of building up the cast thickness which should be completely covered with concrete. Do not paint mild steel when it is used as reinforcement in concrete.
6 Trim the seam edges and make sure the filled caps fit to the main mould. Make any adjustment necessary. The caps and main mould are, of course, filled separately, the order in which this is done depending largely on the nature of the form and mould. I prefer, whenever possible, to fill the caps first, they are then usually ready to be placed without mishap, when the main mould is filled.
7 Fix the caps to the main mould to complete the casting. The caps, and the respective filling, can be joined to the main mould by:

   a Placing the cap and securing it with a clamp, or scrim and plaster. Then pack the concrete, filling the seam on the inside.
   b Squeezing the cap, and filling by placing neat cement, or a mix with a fine aggregate, along the seam edges of the filling of both caps and main mould. Then squeeze the cap to the main mould, forcing out any surplus material.
   c A combination of squeezing and packing.

8 Cover the mould and filling with wet or damp cloths according to the type of cement being used.

   a Aluminous cements must not be allowed to dehydrate during the set, hardening or cure. Wet cloths must be kept on the mould and filling for at least 24 hours. Seal the moisture in during this period by wrapping a plastic sheet around the mould and the cloths.
   b Portland type cements cure by gradual dehydration and oxidisation. The mould and filling must be covered with damp cloths, which should be allowed to slowly dry out over a period of two or three days.

   This covering with wet or damp cloths must be done whenever the filling is left for a long period, on completion of the casting or at any time during the casting.
9 Chip the cast out when the concrete has cured. Make good any flaws in the casting. Trim off the seam flash.

Points to remember

1 There must be an even thickness of casting media over the mould surface.
2 High points in the negative mould are low points on the positive cast, and these can easily become holes if not covered sufficiently in the casting process.
3 Consolidation of material, or proper impregnation of fibre is essential to achieve a consistent, dense, cast thickness.
4 Reinforcement should be well placed within the cast thickness.
5 Neat and appropriate seam edges are important to facilitate proper filling of the seams. Care must be taken in the control of moisture during setting, hardening and cure.
6 Flaws on the surface can be made good by soaking the cast and repairing with the cement used in the casting; this must be cured properly.

Other methods of repairing:

1 Prepare a mixture of the appropriate cement and PVA emulsion. Make this the consistency which enables the repair to be made, it can be liquid or stiff like clay. Prime the surface with dilute PVA and water, then model the repair.
2 On a dry cast it is possible to make a good repair by using polyester resin, filled with the appropriate cement or filler to simulate correct colour. This technique is often the most effective and is certainly the quickest.

Any tools used during the casting processes should be cleaned, when the casting is complete. Brushes in particular, which have been used to apply the first layers of concrete and to stipple the glass fibre, must be washed before the cement sets. Remember aluminous cement sets under water, so sink, drains, bowls and buckets should be thoroughly washed before the set occurs.

Terracotta Press Moulding and Slip Casting

Castings of clay to be fired can be made from plaster of Paris moulds. Indeed, most of the normal household crockery is made by casting with clay from such a mould. There are two basic methods:
A. The first method is a technique of pressing clay into a plaster piece mould. The clay must be of good quality and even in consistency, and is used to build up by pressing an evenly distributed thickness over the mould surface. This technique has been used for many centuries, indeed there are examples of press mould-casting from almost every culture. In the British Museum those from Mesopotamia must be among the earliest examples. Most good archeological museums will have examples of products made using this technique, and a study of such examples will prove valuable as well as interesting. The mould, for very simple things, can easily be made of fired clay, indeed in the past they often were.
B. The second method is to use a liquid clay (slip) which is poured into the prepared mould, filling it. The porous mould sucks moisture from the clay slip, causing a deposit of clay to be evenly distributed on the mould surface. The longer the mould stands filled with slip, the thicker the deposit of clay will be. When the required thickness has been deposited the remaining liquid is poured off.

These two methods can be practised on the piece mould, made from the basic forms; the qualification being that for press moulding the mould must be open ended, to allow the hand access for pressing the clay and for closing the seams, also to consolidate the clay thickness. Obviously for pouring a liquid into a mould, it is necessary to provide a suitable opening. This must be large enough to allow both the pouring in and out, of the clay slip. Practise the two methods of casting with clay on as many of the basic shapes as possible. Add to these shapes an open pan or dish shape on which to practise the actual technique of pressing. The process of filling a mould by this method is as follows:

1 The preparation of the mould is important, as always. This should be bone dry and lightly dusted with French chalk or talc. Do not allow the dust to build up in any hollows in the mould. Old fashioned fire bellows are useful to blow off excess dust.
2 Prepare the clay, enough to complete the job in hand. Wedge or knead it to an even consistency and colour, and make sure it is free from lumps, air pockets and any foreign matter. It should be very plastic but not sticky to handle.
3 Select the mould or pieces of mould to be filled first. An open dish mould can, of course, be filled in one operation. A mould with two or more pieces should be tackled by filling the main mould or large pieces first.
4 Press into the middle of the mould, or piece of

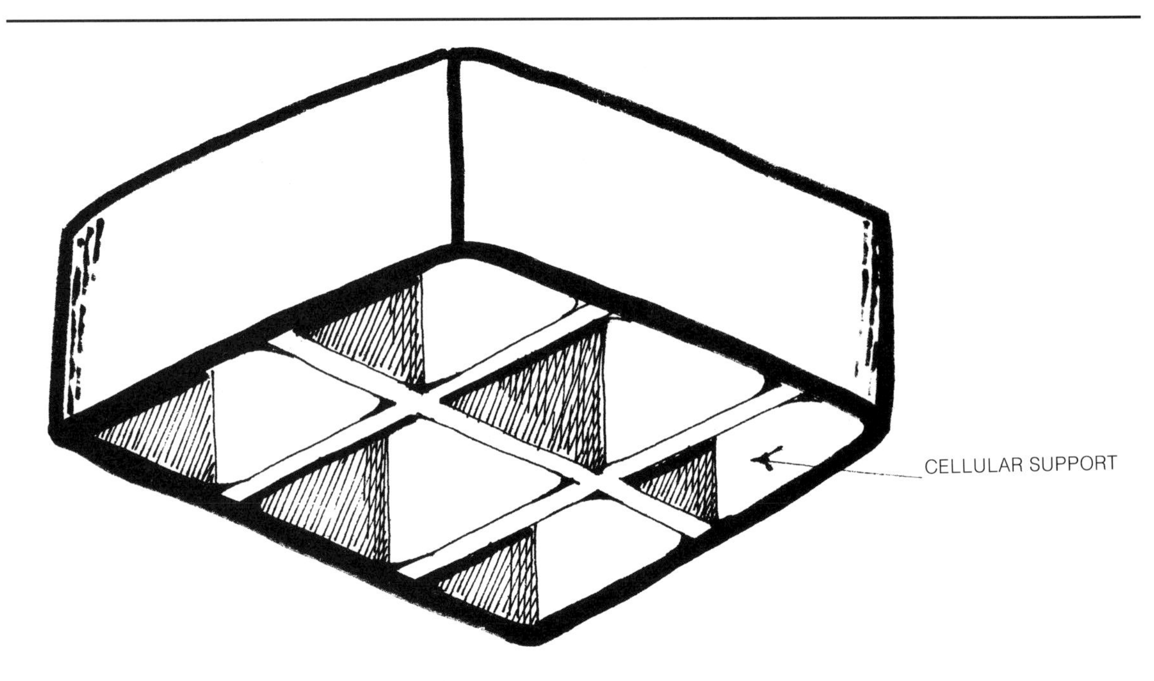

68 Cellular support inside a terracotta form.

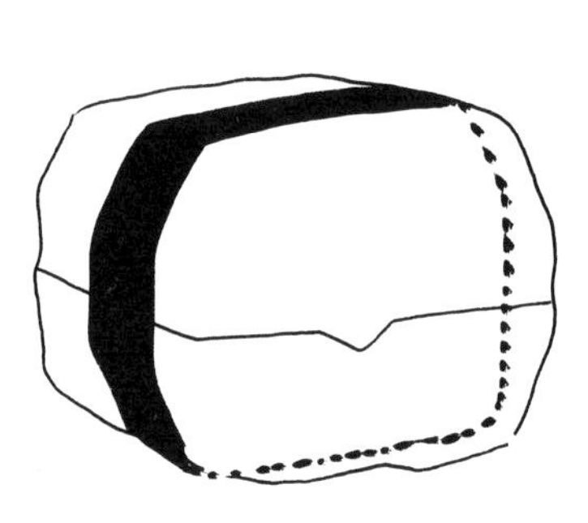

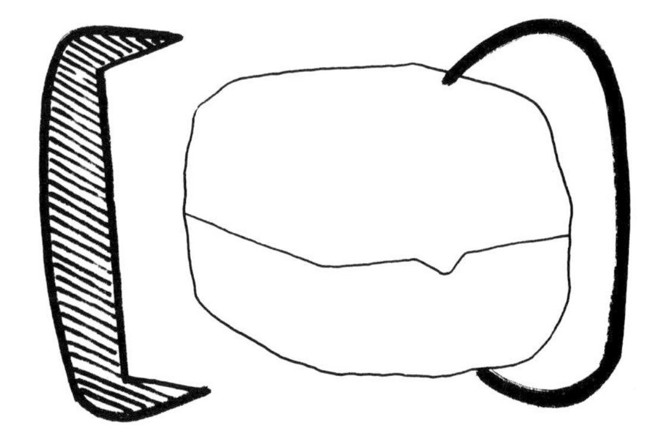

69 Various methods for holding caps in position; useful both on waste moulds and piece moulds. A: secured by means of a thick rubber band (can be cut from an old moter inner tube). B: metal clamps – a carpenter's dog that draws the sections together as it is tapped on. The second is a clamp made easily from a piece of mild steel.

mould, a wedge of clay. Press it out in all directions. Add another wedge of clay to the first and again press outwards in all directions. Similarly with more wedges of clay, press and spread outwards, until the mould is covered evenly over the whole surface. It is important to add clay to clay, and not to add the new wedge beside the clay already on the mould. French chalk from the mould surface should not be allowed to get between the applications of clay. If it does, it will form an effective barrier, and prevent the adhesion of clay to clay. Complete the pressed filling up to the seam edges.

5 Complete the next piece or pieces of mould in the same way. If the cast requires strengthening, ribs of clay must be placed to provide this. They should be placed during the build up. A honeycomb or cellular structure is the most suitable for clay forms, affording maximum strength. The amount of cellular support depends upon the volume of the form, and consequently on the area of surface to be supported.
6 Place completed caps to the main mould, first wetting the clay at the seams to make a better adhesion. Fill the seam and strengthen from inside the form. Place and fill the seams of each cap in the necessary order to complete the form. The caps when placed to the main mould can be secured by means of mild steel clamps. An open dish obviously will not require more work than is described in paragraph 4.
7 Having completed the filling, stand the mould upright if possible on an open shelf, upon battens of wood to allow air to circulate. Cover the mould with newspaper or a light fabric and allow the clay filling to become leather hard. This may take a matter of a few days to achieve. Do not hasten the drying artificially. When the clay is leather hard, carefully remove the mould. It will be noticed that the filling shrinks from the mould, making it easier to remove.
8 The casting is now ready to be cleaned off and prepared for the final image, i.e. assembled, painted or glazed.

Slip casting is a much quicker process, in the initial stages, than press moulding; the procedure is as follows:

1 Prepare the mould, which must be dry and lightly dusted with French chalk. If the mould has more than one piece and is without a piece mould jacket, the pieces must be secured together. This can be done with thick rubber bands which can be cut from an automobile inner tube.
2 Prepare the slip to be used in the casting; i.e. have enough to fill the mould at least. Clay slip can be made from any fine grained clay by adding water. Sodium silicate or soda ash added in the proportion of $\frac{1}{2}$ per cent will help to make a clay slip quickly. More complex bodies for slip casting can be used, but it is wise to seek the advice of a ceramics artist who is familiar with the material available in particular areas.
3 Pour the prepared slip slowly and smoothly into the mould. Vibrate the mould gently to release any air bubbles.
4 Leave the filled mould long enough to deposit the required thickness on the mould surface. Then pour off the excess fluid.
5 Leave the mould and cast on an open shelf, supported on wood battens to allow air to circulate. Stand the mould upright if possible. Cover the mould with newspaper or a light fabric, to make the drying out process slow and even. Remove the mould when the clay filling is leather hard. It will have shrunk from the mould and will therefore be easy to remove.
6 The casting can now be cleaned off and prepared for the final treatment of colouring, glazing, painting or assembling.

Important points to watch:
1 The amount, consistency and purity of the clay for pressing, and slip for casting.
2 The even thickness of the material over the mould surface.
3 The proper strengthening of a clay press cast. These forms on the whole can be larger than slip cast forms.
4 The castings must be evenly and gradually dried, in the mould and after. It is best to do this in a controlled warm atmosphere, (not hot). Do not artificially dry the cast as this may lead to uneven shrinkage and cracking.

Wax

Wax castings can be made from plaster piece moulds. Indeed this was common practice in foundry work, before the introduction of flexible materials for moulding. The wax most commonly used today is micro-crystalline wax, a by-product from the refining of petroleum. This is a most useful material, widely used both for casting forms and for modelling direct. Making a casting with this substance from a plaster piece mould is similar in technique to a combination of press casting and slip casting. No release agent is necessary but the mould must be saturated with cold water. Wax resists water and vice versa, therefore no adhesion can occur when wax is applied to a wet mould surface. The application of the wax is similar to the application of clay in a press casting. Wax, however, is applied with a brush, and must overlap. Make a brush mark, then the next, brushing from one mark to the mould. A soft, full brush is the most suitable for this purpose. The idea is to cover the mould surface with hot wax by adding a carefully applied layer of wax. Over this paint, fairly freely, a second layer, building a more substantial thickness at protruding forms on the mould surface. Then trim the seam edges and put the mould pieces together if there is more than one piece. Secure the mould and pour in molten wax, which has chilled a little. In this way a deposit of wax is made to increase the painted application and to make the thickness required. Excess wax is then poured off when the cast thickness is reached.

The preparation of the wax is an important factor. It can be bought in pre-determined hardness and softness. If a wax needs to be made softer, however, this can be achieved by adding to molten wax a proportion of pure petroleum jelly, up to 25 per cent; alternatively a heavy engine oil can be added, in the same proportion. To harden a too-soft wax add a small amount of amber resin, but not more than 20 per cent, or the wax will become very brittle. The wax can be melted in a metal vessel, a saucepan, bucket or small galvanised bath. The vessel used depends largely on the volume of form to be dealt with. Place this over a low heat. Some people prefer to cushion the vessel in which the wax is melted, by placing it in a second vessel containing water, which slows the melting of the wax considerably. This method though slower is safer, as large amounts of wax, when melted over a direct heat, melt from the bottom, and can build up pockets of air, which expand to eventually burst through the cooler upper levels of wax. This causes the molten wax to spurt, which can be dangerous, especially in a class or studio. To avoid this, either cushion the vessel in which the wax is melting, or push a heated metal rod through the wax to make an airway for expanding pressure to escape. Cushioning the melting vessel causes the wax to melt from the sides and avoids a pressure build up. *Be very careful of molten wax*, it can cause a nasty burn.

The procedure for making a wax casting from a plaster of Paris piece mould is as follows:

1 Prepare the wax, melt it and allow it to cool till just before it becomes cloudy on the surface. This will become familiar as experience of melting and using wax is gained. A small amount can be allowed to cool in a separate smaller vessel, while keeping the larger amount molten and hot to draw from.
2 Prepare the mould whilst preparing the wax. The mould needs simply to be wet.
3 Place the first application of wax to the surface of the mould. Do this with a fully charged brush, placing it simply to deposit the wax on the mould. Do not try to paint this application, because the wax will chill on the brush and drag, pulling itself from the surface. Cover the entire mould surface evenly with wax applied in this manner.
4 Paint a second application of wax. This

layer can be painted quickly. Try not to let the brush stick and drag. Keep dipping the brush into the wax to make it spread.

5 Using some chill wax, from the side of the vessel or from another pot, build up and thicken high points and protrusions on the mould surface, remembering that these will be hollows or low points on the cast. The reason for building them up in this way is to ensure they maintain their coverage when molten wax is poured into and then out of the mould.
6 Clean off at the seam and trim along the seam edge with a sharp knife.
7 Place the pieces of mould together, secure them with rubber bands or with string and wooden wedges.
8 Pour molten wax into the mould. The wax must be poured when it has cooled and the surface has become cloudy. If used before this it will be too hot.
9 Allow the filled mould to stand awhile, so that a chilled deposit of wax is made on the surface of the mould. When this deposit is the thickness required, $\frac{1}{8}$in (3mm) to $\frac{3}{16}$in (4mm) pour off the surplus wax. Then allow the casting to cool and harden in the mould. This process can be accelerated by pouring cold water into the casting. (It is at this point that the core of refractory material can be placed, to prepare the wax casting to be invested and made ready for bronze casting, see page 125.)
10 Carefully remove the mould when the wax is hard. Clean the casting, make good any defects by adding molten wax, and modelling with a heated metal spatula. Remove any seam flashes and prepare the casting for whatever its final state is to be; the intermediary stage of a metal casting, or a wax final material of the intended image.

Points to remember

1 A wet mould.
2 Wax at the right temperature.
3 Careful application of the first layer of wax.
4 Sensible build up of the high points on the mould surface.
5 Pouring the molten wax at the correct temperature; not too hot to melt the painted layers, not too cool to chill faster than is controllable.

I will deal later in the book with the technique of cire perdue (lost wax) bronze casting for which the above description is a very vital preparation. The history of lost wax casting is a very long one and the important part wax plays in this is of course equally long. The chapter on metal casting will deal with this aspect more fully.

Synthetic Resins and Polyester Resins

The basic method employed in casting with polyester resins, which are probably the most widely used synthetic resins in sculpture, is similar to most other materials. Differences, however, arise from the peculiarity of the material, demanding some understanding of the various chemical changes that take place during the cure. The mould required is similar to that made for packing with concrete or clay, it must offer maximum access to the mould surface. The state of the mould to be prepared must be dry, as with casting in clay. The release agent is quite unique, however, since it has to deal with the difficult task of releasing an extremely persistent adhesive material.

This material is at once a complex chemical substance whilst being fairly simple to use. It is usually made up by mixing three proprietary liquids; (1) the basic resin to which a chosen filler may be added; (2) the promoter or accelerator which controls the curing speed of the basic resin, and is added to hasten the chemical change from a liquid to a solid; (3) the catalyst, which is the chemical that, when added to the basic resin, causes the chemical change, resulting in a hard insoluble material.

The resin, accelerator and catalyst are usually sold under these headings. Resin suppliers will advise the appropriate materials, and additives of all kinds, compatible with their products. Be certain to ask for the accelerator and catalyst to use with particular resins. Do not rely on samples from other manufacturers. These additives do vary according to the nature and origin of the raw material, from which the resin is made.

It is important to use these materials in the order in which I have mentioned them. This corresponds to the order of mixing:

1 Filler into resin.
2 Accelerator to resin.
3 Add catalyst as and when required to harden the resin.

It is wise to mix the total quantity of resin and filler needed for the work in hand, to maintain consistent colour and cure. It is possible to add the accelerator to the whole resin mix, to attain a constant rate of curing throughout the work. Do NOT add accelerator to catalyst; such a mixture can react violently or explosively. It is wise to determine the length of time the pre-mixed resin-accelerator takes to harden, after adding the catalyst. Then add the catalyst to make sufficient resin to complete a determined area of application easily, allowing time to clean brushes and other tools before the resin gels and hardens. The cleaning of tools must be done diligently after each fresh mix. This cleaning can be done by washing the implements in warm water and detergent, or in an appropriate solvent for the particular resin used. Mixing the resin in a large amount, and taking smaller quantities from that to work with, saves energy and cost, which would otherwise be spent on making smaller amounts more often. It also ensures an even rate of cure throughout the work, which in the long run is best. Difficult areas can easily be dealt with by adding extra accelerator to a small mix, to speed up the gelation and cure. This method is often employed to secure reinforcement, or to fill a tricky overhanging surface.

The proportions of the components is important, and the manufacturers' advice should be sought at the time of purchasing the resin from the supplier. Generally the proportion of accelerator should not exceed 4 per cent. More than this proportion may, with some resins, act as a retardant. The amount of catalyst should not exceed 3 per cent. More than this does not increase the speed of cure, and is therefore a waste of material.

Fine detail reproduction required from a casting medium is achieved by painting on the surface of the mould a layer of resin known as the 'gel coat', whose job it is to make the faithful reproduction of detail. This gel coat must be applied with care and should be as even as possible. Any colouring or filler used in the resin should be well mixed and dense, especially for the gel coat. This is particularly important when using a metal filler. Uneven mixing will possibly lead to a patchy uneven colour on the final surface, as indeed will an unevenly applied gel coat.

Resin, when hard, is fairly brittle and therefore requires some form of reinforcement. To achieve a strong material, laminates of resin and glass fibre are built to make the required thickness. This allows an efficient tensile strength. A mat of glass fibre is stippled on to the hardened gel coat with resin. Stippling ensures that the fibres of the material are properly impregnated. It also helps to drive the fibre into the finest forms, to make sure they are strong. This process of lamination is called 'laying up'. It is important to make every fibre bond well to ensure a good dense consistent thickness. Tools of various and curious shapes are made to assist in placing the fibres where they are most needed, to force them into difficult forms, and help an even 'lay up'.

Other kinds of reinforcement may be included in the laminates to give greater rigidity and strength to a casting, such as mild

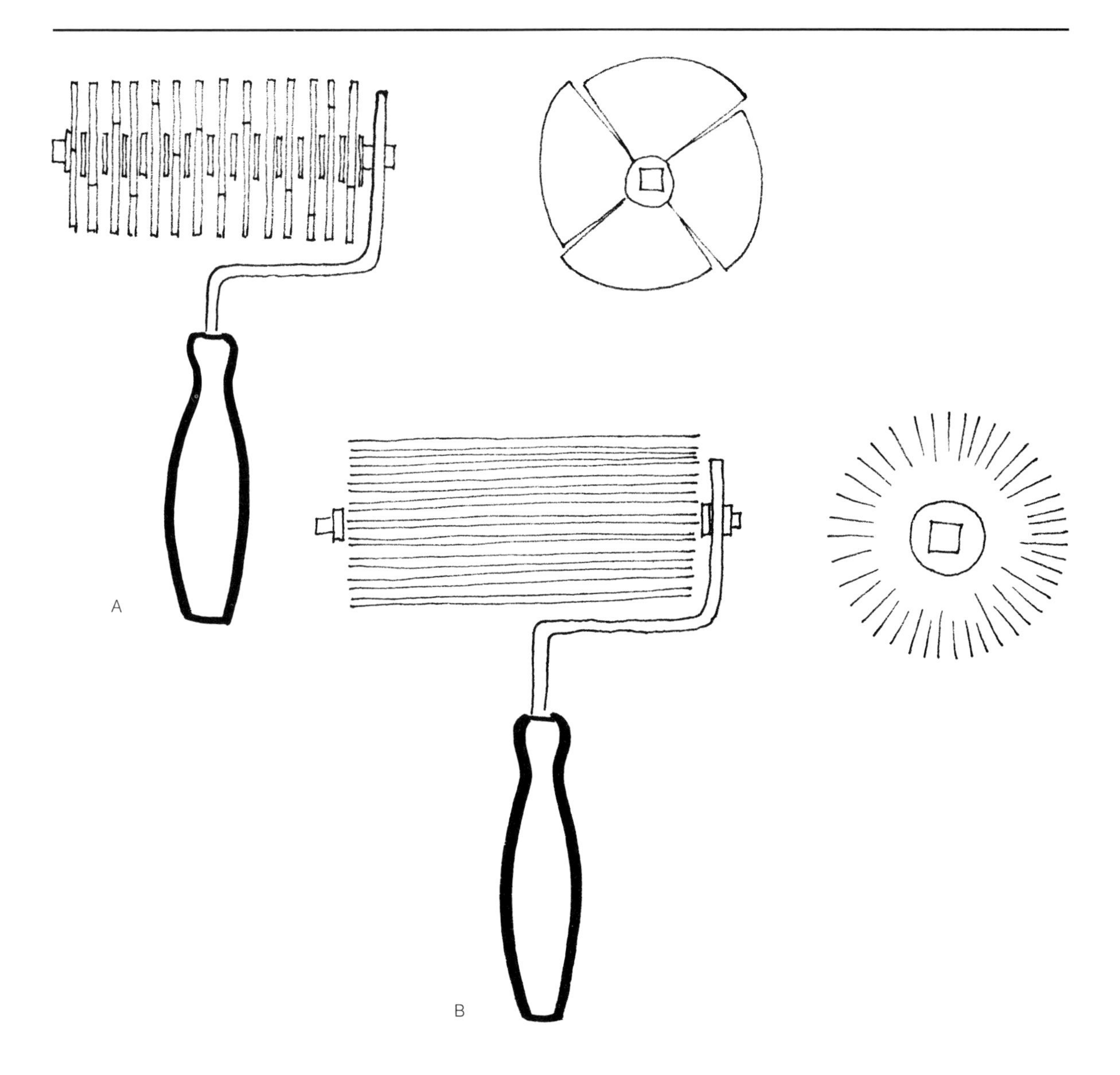

70 A series of rollers used to make good laminations of polyester resin and glass fibre. A: roller made from split washers. B: roller with horizontal separation, useful when laying up on broad areas. These rollers and variants can be made up in the studio to fit particular jobs.

steel, high tensile aluminium, wood, and nylon rope, in fact almost anything inert. It must be included in the 'lay up', and be covered completely with resin and glass fibre, which will provide proper protection from any corrosive attack.

When the resin has hardened and cured, the mould can be removed. In this chapter it will mean chipping away the plaster waste mould, or in the case of piece moulds, carefully prising the mould off. Flexible moulds will be explained later in the book.

The nature of the material, of course, largely determines the method and usability of that substance. Any misunderstanding of what, in general, is happening to the media at given moments, contributes to a misuse or a failure. Good handling and proper appreciation of the use and suitability of form and method result from an understanding of the material.

Polyester resins are unsaturated resins, which means they are capable of becoming hard solids under certain conditions. They are thermosetting, becoming hard and inflexible by the heat generated during their cure (exotherm), which is brought about by adding the catalyst. The term polyester is derived from polymerised esters. Polymerisation is a non-reversible chemical reaction, and is employed to obtain polyesters by reacting dibasic acids and glycols. At this stage the molecular structure of the resin is in the form of a chain of identical molecules, simple repeating units. The polyesters are dissolved to form a solution, in a reactive monomer. This monomer enables the chain of esters to cross link, after the addition of a catalyst, to form a complex three dimensional molecular structure. This is in fact a further polymerisation resulting in gelation and cure of the resin, to become an inflexible and insoluble material.

Some of the raw materials used in the manufacture of polyester resins are as follows:

DIBASIC ACIDS	Maleic
	Fumaric
GLYCOLS	Ethylene Glycol
	Diethylene Glycol
	Triethylene Glycol
MONOMERS	Styrene
	Methyl Methacrylate
CATALYST	Benzol Peroxide
	Methyl Ethylketone Peroxide
	(Low temp.: < 15°C (59°F))
ACCELERATOR	Cobalt Naphthanate
	Di-methy Aniline

When the three components – resin, accelerator and catalyst – are mixed there begins the process of polymerisation. Visible changes take place, and the resin changes from a syrupy liquid to a jelly-like solid, which in turn becomes very hard when cured. Resins will slowly polymerise without the addition of accelerator, or catalyst, and this factor largely determines the pot life, or shelf life, of a particular resin, i.e. the length of time it can be stored. This polymerisation may take place over a period of years, but should be watched, and the resin stored in a cool, dark place. Do not store resin in glass jars, if exposed to sunlight; in such containers the shelf life will be drastically reduced, sometimes down to a matter of days.

This explanation is very general and basic, and is meant to give an idea of the medium and what is going on inside the mould when the casting is being made. There are of course things that can go wrong, and it is useful to have some idea of what should be happening, and so recognise, and if possible overcome or avoid such difficulties.

A large contributory factor in the efficient cure of a resin is heat. Exothermic heat is produced during polymerisation of a resin, and this heat enables the resin to cure. The temperature of the studio in which the resin is being used is important too. Resins are generally made to cure in a controlled temperature of about 15°C (59°F) to 150°C (302°F), and if the working temperature is below 15°C (59°F) there is a risk that the resin will not cure. A low temperature retards the cure drastically, with the serious effect of causing the monomer to evaporate, spoiling the cross-linking polymerisation, resulting in a sticky jam-like surface. To avoid this, a resin that will cure at a low working temperature should be used, resin manufacturers will advise on this. Currents of air will also cause the styrene monomer to evaporate, even if it is warm air. An application of heat externally, often helps to further accelerate a cure. Industrial infra-red heater units are useful in this repect; they will also help to raise the working temperature in the area of the mould.

Exothermic heat can be very damaging and may catch off guard the unwary sculptor. Resin, if mixed and left in a large volume, will create great exothermic heat, which will cause the material to expand and contract violently, resulting in large cracks. For instance, a general purpose resin poured into a cubic container, and allowed to cure at around 15°C (59°F), will generate a temperature of up to 160°C (320°F)during gelation and hardening. The thicker the casting the greater the heat. The use of lamination and the introduction of fibres and fillers of various kinds, helps to avoid this hazard of excessive exothermic heat.

Fillers are used in polyesters for a variety of reasons. They are added to reduce exotherm, i.e. heat. They are also added to achieve particular colour and effect, and to reduce cost by increasing volume (fillers are less expensive in most cases than resin).

Fillers are also added to provide compressive strength; fibrous fillers give additional tensile strength. Polyester resins are usually of pale transparent colours which in themselves are very beautiful. This transparency can be avoided by the addition of fillers or pigments. It is wise to experiment with any intended filler, however, because some can prove to be efficient retarding agents, and can even inhibit the resin cure.

A filler can be useful in assisting a lay up, by making the resin thixotropic, and therefore making it less likely to drain away from an inclined or vertical surface. The limitation of the amount of filler that can be added is determined by the ease with which a mixture can be used.

71 Glass fibre reinforced polyester resin sculpture by Richard Rome, postgraduate student, Chelsea School of Art.

72 Mould for the upright section of the sculpture made of the same material.

73, 74 and **75** Polyester resin and glass fibre laminates being built up on the mould surface to make the casting. The 'gel coat' has been applied first and allowed to harden, the subsequent layers are built up by stippling the resin to the glass fibre to impregnate it.

76 A cap is placed to the main mould.

Fillers frequently used in resins include the following:

Alumina	Slate Powder
Calcined Clays	Whitening
China Clays	Wood Flour
Metal Powders	Talc
Mica Powder	Glass Fibre
Aerosil (to gain thixotropic properties)	

These fillers are added as powder or fibres; they must be evenly mixed and the fibres well impregnated with resin. They may, however, affect the curing rate of the material by diluting the resin, thus reducing exotherm, and prolonging the cure. Or they may increase exotherm, accelerating gelation and cure. Before making large quantities of mixed filler and resin, test the reaction.

Pigments for colouring resins can be bought in paste or powder form. Paste is easiest to use. Powdered pigment must be thoroughly mixed to a paste with a small amount of resin prior to mixing the larger quantity. In this way the best distribution of colour is ensured.

Use the pigments recommended by the manufacturers. Pigments used in other industries may not be compatible to polyester resins, and they may inhibit gelation and cure. Experiment wisely with any new powder.

The procedure for making a casting with polyester resins is as follows (this description is the method employed in all the basic forms, and is in fact typical in general to all casting with this material):

1 Prepare the mould, which should be dry. See release agents, page 52.
2 Mix the large quantity of resin and filler, plus accelerator. Prepare enough to complete the work.
3 From the large quantity take a smaller amount, mix in the catalyst necessary, and apply this to the surface of the mould. This is the 'gel coat'. Add a thixotropic agent, if the mould has inclined or vertical surfaces; this will prevent the resin draining and running off. Make the gel coat as even as possible. Clean any brushes immediately after application, before any particular mixes gelate.

   Apply a second coat of resin to build up the gel coat, if the texture on the surface is deep. A substantial gel coat is an advantage if work is to be done on the surface of the cast.
4 Cut the glass fibre to fit the mould with a little overlap.
5 Mix another small amount of resin and catalyst. With this make the first lamination by stippling the glass fibre to the gel coat, when the latter has hardened. Build up the laminations to the required thickness – ¼in (6mm) maximum. See figure *63D* for diagram of layers of lamination.
6 Fix any other reinforcement during the laminating lay up.
7 The seam edges should be kept clean and any fibres turned back into the lamination or trimmed off.
8 The seam edges can be built up by using a thixotropic mix. The shape at the seam edges and faces should be as in figure *60I*.
9 Treat the caps and main mould as described above until the whole mould surface has been cast, the laminations built up to the required thickness, and all seam edges and surfaces trimmed and prepared, and are ready to be fixed together. Ensure that all the caps fit snugly to the main mould, and to each other. Trim with a file or hacksaw any protrusions which prevent a close fit.
10 Fix the caps, one by one, to the main mould. Use the squeeze method, securing the caps in position by means of clamps. Then reinforce each seam on the inside, with glass fibre and resin, whenever possible. Allow the resin to finally gel and cure when the final cap has been placed.
11 Chip off the waste mould. This can be removed more easily if it is soaked first, then chipped off carefully. Piece moulds should be carefully prised off, trying not to chip and damage the pieces. These should be taken from the cast in the planned order, and replaced in the mould case. (The piece mould for resin, when of plaster of Paris, is best made from a low expansion hardened plaster.)

When the casting has been released from the mould, it can be cleaned off and any defects to the casting made good. The same resin mix must be used for making good the cast, and if this is done carefully the repair will not be seen. Prior to making good a defect, be certain to clean the surface of the casting with steel wool or a wire brush. This is essential to ensure a proper adhesion between the resins, which cannot be achieved if any release agent

77 The waste mould is chipped off to reveal the casting.

remains on the surface.

Small moulds can be filled by pouring. This is done by painting a gel coat carefully, then closing the mould and securing the pieces. Into this it is possible to pour or pack what is known as a dough mix. This is a mixture made using a low viscosity resin, filled with both a powdered and fibrous filler, which increases the volume of the mix, and at the same time reduces exotherm, permitting a small mould to be filled solid. The rate of cure of such a dough mix must be slow and carried out in a compatible, consistent, temperature.

Castings should be carried out on many basic forms so as to become familiar with as many techniques as possible before undertaking a more important sculpture.

Points to watch during the filling

1 A proper preparation of the mould.
2 That the release agent is well applied.
3 Avoid an uneven gel coat, due to poor mixing or haphazard application.
4 Avoid air bubbles being trapped between the laminates; these must be released during the laminating process. If this is not done, the cast surface is liable to be weak and pitted with large holes.

Careful reinforcement must be sensibly placed to ensure tensile strength. Moisture on the brushes or tools will inhibit resin cure; to avoid this use several brushes, allowing them to dry

78–81 Strong reinforcement being placed to strengthen a glass fibre polyester resin sculpture to be situated in a public site. Heavy-gauge galvanized pipe is used, running the length of the figure and into the plinth (*Opposite top illustrations*). The tubes are cross braced and secured with resin and glass fibre (*opposite bottom*). The plinth is finally filled with concrete to secure the sculpture (*above*).

properly before re-use. Do not use brushes straight from a solvent. Do not allow draughts to blow across the work, or the temperature to drop below the optimum working temperature, for the particular resin in use. The styrene monomer must not be allowed to evaporate. Clean the tools and brushes before resin sets on them.

Casting with Polyurethane

Polyurethane is most widely used as a flexible foam material for upholstery and packing. It can, however, be bought as a two part castable synthetic resin comprising the resin on one part and the catalyst on the other. When the two liquids are mixed together, the reaction caused by the catalyst makes the material foam and expand (2lb (900g) of liquid will expand to 2cu ft (60cu m) of foam). This expansion occurs quickly and exerts great pressure during the process. The resulting solid is of a fine honeycomb structure with a dense surface skin. It is very lightweight and buoyant and can be used as a filler to provide lightweight strength for hollow casting such as those made with GRP or concrete. It can also be used as a solid form or cast to be carved and constructed to make a sculpture. There are proprietory adhesives available that can be used for all these methods. Equally, it can be used to make cast and carved moulds or shuttering, from which other castings can be made. With this technique large architectural decorations can be made, often cast in situ. The mould can be removed by prising it off, dissolving it with suitable solvents or burning it off; the last technique will result in isocyanitic gases so should only be resorted to as a last resort and only in the open air.

Because of the pressure exerted as the resin expands and sets, any mould must be adequately reinforced and sections securely clamped together to resist the pressure. The poundage exerted by the foaming process can be up to at least one ton per square inch (50kg per square millimetre), and so a completely enclosed mould must be strong enough to resist such pressure). Such an enclosed mould however is only necessary to gain the most dense foamed solid. It is wise under all other circumstances to provide a small pressure escape vent to relieve stresses on the mould during the expansion and set of the polyurethane. The same precaution should be taken when using this foaming resin as a filler. Try to control the effect of the expansion so that no damage is caused to the object being filled because any distortions that are caused by the expansion are very difficult to correct.

Polyurethane foam has fine adhesive qualities, and is often used as a sprayed insulation material for buildings. It will therefore require a very effective release agent to allow it to be easily removed from the mould. Water prevents it sticking and so porous moulds can be filled wet. Other moulds need to be suitably sealed and then treated with an oil release agent. Some experiments will be necessary to provide you with a satisfactory result. Consult the supplier or manufacturer of the material who may be able to provide a proprietary release agent specifically designed for the chosen product.

Polyurethanes are very toxic materials and so any work space in which they are used must be well ventilated and breathing must be protected using a suitable mask.

III Plastics

Only relatively recently have plastics found their way into sculptors' studios. This has largely been brought about by the advent of cold curing resins known as RTVs (room temperature vulcanising). The term 'plastics' must be regarded simply as the generic heading under which all synthetic resins are grouped. The term derives from the fact that these materials, at some stage, are soft and pliable enough to make possible modelling and moulding, later becoming firm and hard to retain the modelled or moulded form. Other materials in the sculptor's repertoire have similar plastic properties, of course, such as clay and plaster, but these have come to be excluded from the group of plastic materials because of the present-day fact of the plastics industry and synthetic resins or polymers. The infinite number of resins manufactured vary in kind, formula and application, and would be an all-consuming study for any one man to keep abreast of them all. Selections, therefore must be made according to how the resins handle and the result they give.

The plastics used in the sculptor's studio, apart from the PVC sheet, which is used to wrap up clay to keep it wet, are fairly few, considering the large number on the market, the greater proportion of which are machined plastic, in the sense that they require complex machinery to put them to use, to produce any form. The principles upon which some of the machines work I have indicated in Chapter VI. The sculptor's choice is therefore the basically thermosetting resins which are dependent upon exothermic and externally applied heat to make them hard and insoluble. Thermoplastic resins are little used in the studio, but are used in industry to produce everyday articles in limitless number. Cold curing resins are those most commonly used by sculptors, the polyester group being the most widely favoured. Acrylic resins, the qualities of which are still not exploited fully, are used to some extent, as are polystyrene and epoxy resins. Polyvinyl chloride is used in the form of PVC sheet, and it is also used as a moulding compound (thermoplastic). I know of no other plastics commonly used by sculptors, except for buckets and bowls and other items of equipment manufactured by the industry.

Polyester Resins

As moulding materials polyester resins as just as valuable as they are as casting media. The hardness of the set material together with its lightweight character, are factors in its favour. The variety of integral colour that can be achieved is also of great importance, a factor which has contributed in some part to polychromatic sculpture becoming so widespread among the younger sculptors. Polyesters can be made to simulate metals and this is the quality that first endeared these resins to the world.

As a moulding medium polyesters have a particular application. They are used to make precise castings of polyester glass fibre laminates, particularly in the motor racing busi-

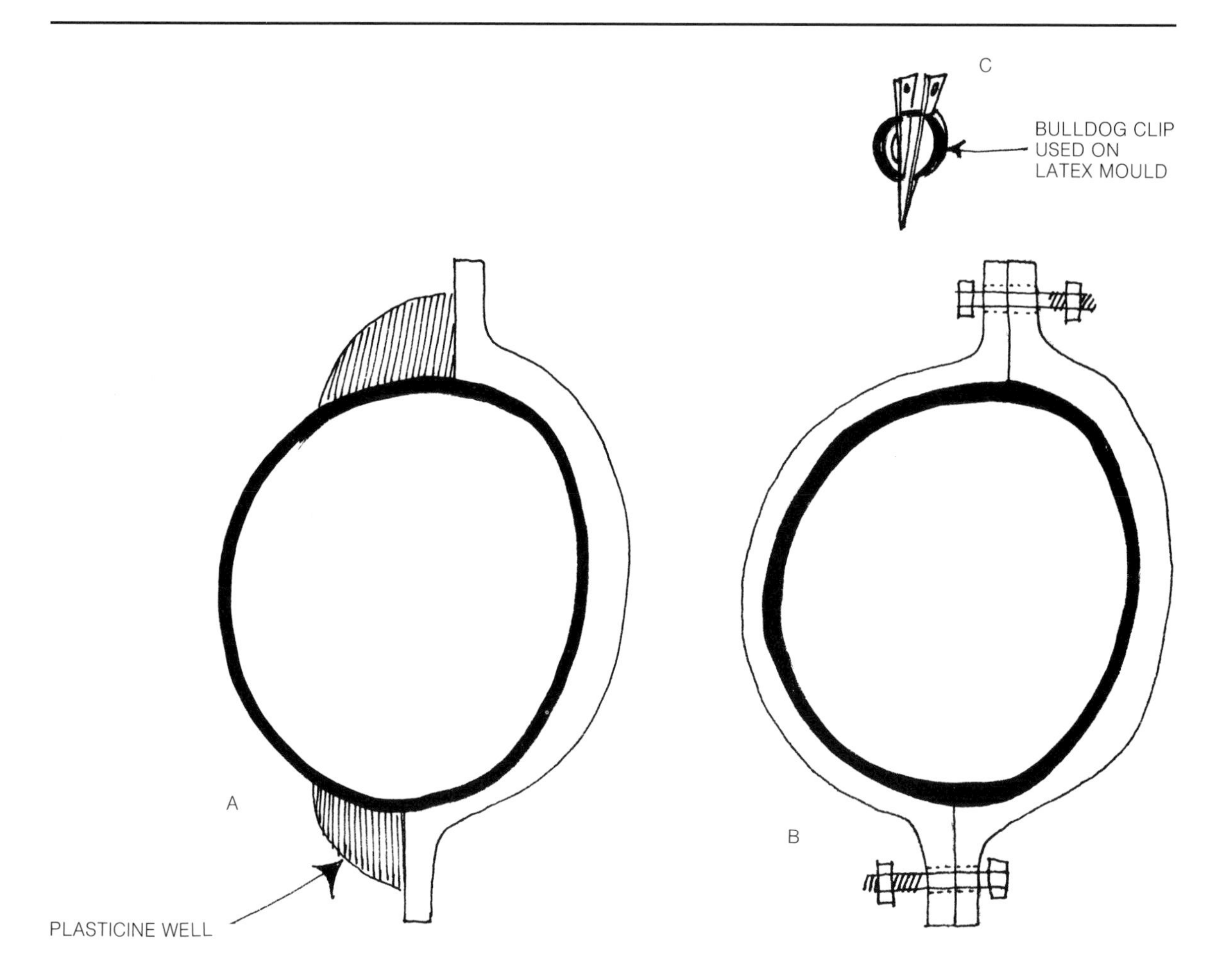

82 Seam surfaces and fixing polyester mould. A: the moulding principle used to make firm, well-registered seam surfaces, on polyester resin moulds and cold curing rubber moulds. B: nuts and bolts to secure mould of polyester resin. C: bulldog clip to hold the flexible mould securely.

ness. I have already dealt with the character of the resin, mixing, polymerisation and laminating, in the previous chapter. The kind of mould made with polyesters is a piece mould (52), the only difference being in the way that the seam wall is made. Plasticine is used to determine the wall on the master form. This can be done with some precision. The master is treated with a sealer and parting agent. The pieces of mould are made by building a lamination consisting of: gel coat (fairly thick), resin and glass tissue, resin and glass fibre, plus any reinforcing required. The seams can be drilled and bolts provided to make the strongest possible closure of the mould (*82*). The mould when finally removed from the master cast can be handled freely because of its strength. The surface can be worked on, polished and textured. The release agent is then applied to prepare the mould for casting.

The procedure for casting is as follows:

1 Prepare the master cast, sealer and parting agent. Plan the sections of mould and draw the seams.
2 With Plasticine make the precise wall surface to determine the first section of mould.
3 Make the piece by means of the normal lay

up; gel coat, subsequent laminations with glass tissue, glass fibre reinforcing. Be certain to make the seam edges as in figure *82*.

4 Remove the Plasticine, apply a release agent to the resin seam. Make the next section of mould.
5 Repeat the laminating, Plasticine wall, and release agent processes until the master cast is covered with mould section.
6 Drill at the seam shape to allow for the bolt as shown in figure *82*. The bolts clamp the mould sections together, and these are held firmly in position because of the nature and breadth of the seam surfaces. No mould case is necessary.
7 Remove the mould from the master cast. Prepare the mould surface.

This description is brief, but if taken in conjunction with the section on piece moulding with plaster of Paris, and the section on casting with polyester resins, the principle, and subsequently the practice of mould making with this material will be clear.

Polyvinyl Chloride and Gelatine

After polyesters the next most common synthetic material used in the sculptor's studio, in connection with casting techniques, is a PVC compound made with a plasticiser. This is a flexible material used for moulding, and marketed under various proprietory names, such as *Vina Mould*. It is a thermoplastic material, becoming a liquid when heated, and cooling to a rubber-like consistency; it is most practical for the manufacture of flexible moulds. Such moulds have obvious advantages, permitting a number of identical forms to be made from the one mould, which, because of its elasticity can accommodate the deepest undercut easily. This avoids the time-consuming process of piece moulding, the alternative for producing a number of castings.

The traditional material usurped by the plasticised PVC is natural gelatine. It can be used in most of the same instances as PVC. After use the PVC can be cut up, returned to the pot and made liquid again for re-use. There is no wastage, an advantage over gelatine, which has a larger proportion of wastage, although the initial cost is smaller.

The temperatures at which the hot melted polyvinyl compounds become plastic enough to pour from a melting pot into a mould, are from 120°C (248°F) to 170°C (338°F) according to the grade. Various grades of softness melt at various temperatures: the softer grades melt at the lower temperatures. These temperatures determine the nature of the material of the master cast, which must be able to withstand such temperatures.

Gelatine permits castings to be made only of cold setting material or, at best, very low temperature liquids. This fact, therefore, limits castings to plaster of Paris, and the various types of low temperature wax, which chill quickly. Polyvinyl chloride, because of its high temperature melting point, and its resistance to acid attack, enables a much wider range of casting media to be made. It is possible to use plaster, wax, cement, concrete and various resins, without damage to the mould. No release agent is necessary and the material can be supplied in several degrees of hardness, enabling the most suitable moulds to be made for particular casting processes.

Because of the comparative high melting temperatures it is best to use specially designed equipment to make a pouring of liquid. The cost of such equipment is quite high initially, but because of the wide use of the moulding media the initial outlay becomes an investment quickly returned. An air-cushioned aluminium pot is required to enable the high temperatures to be reached without burning the material. Such a pot can be bought from a manufacturer or constructed by or to the instruction of the sculptor. The hot melt

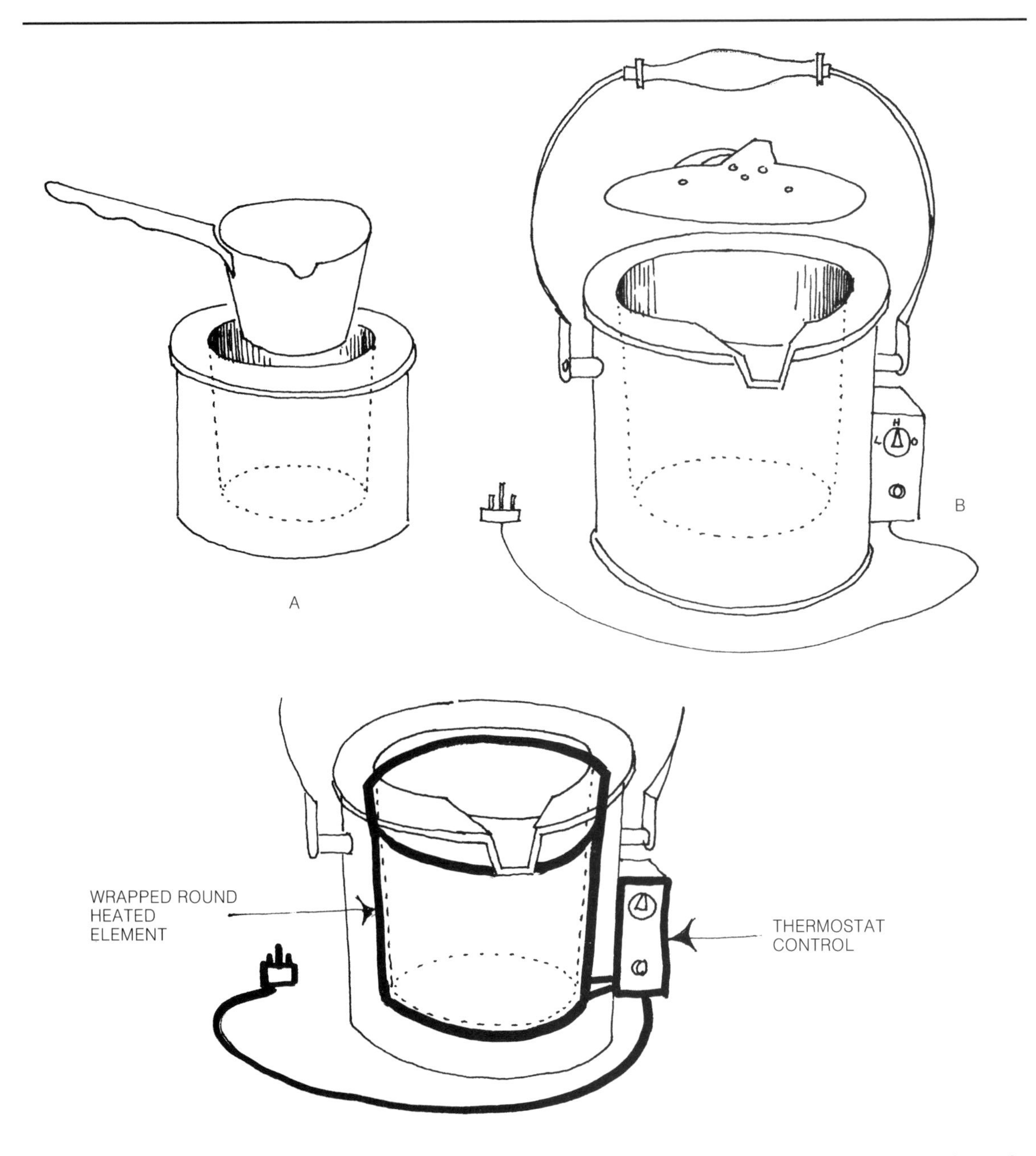

83 PVC melting equipment. A: a simple melting system. B: an electrically heated and controlled melting pot. C: the heating element placed about the container and air cushioned.

compound can be melted in an open saucepan but in very small quantities. This will prove to be a very smelly and worrying business because the substance will scorch and give off very objectionable fumes. An aluminium pot built on the same principle as a glue porringer is ideal, but should not be cushioned with water. A satisfactory cushion can be effected by putting oil between the melting vessel and its container; this will allow the necessary high temperatures to be reached but oil can also give off unpleasant fumes. The space between the vessel and container need only contain air but should be well sealed. A melting pot can also be made from two tins, one larger than the other. A hole cut in the larger tin to house the smaller, makes the pot suitable for melting quantities of up to 10lb (4.5kg) of hot melt compound. Quantities over this should be melted in a vessel electrically heated and with thermostatic control. The material is cut into small pieces before melting and should be constantly stirred whilst melting.

The vessel needed to melt gelatine is simply a bucket inside a larger bucket, cushioned with water. This vessel can be heated over an electric heat source, over a gas flame or a fire. The gelatine should also be cut into small cubes and be stirred well whilst melting. This material is cut into cubes in the first place to allow it to dry properly; it is then laid out on a tray to dry without deterioration.

It is possible to make many different types of mould with the hot melted PVC; these vary from the elementary to the complex. An elementary mould can be made by pouring the molten substance over a shape or form held within a vessel or container; the molten material covers the form and fills the container, and when cool a satisfactory mould is made. Complex moulds dealing with intricate and complicated forms, however, require more involved and specific moulds. One of the biggest uses of this material is in the film industry, in the production of large set pieces.

The basic forms suggested here, on which to practise the various techniques, offer no difficulty in making polyvinyl flexible moulds.

The simplest mould can be made as follows:

1 Commence to melt PVC
2 Place the form on a non-porous surface. An old marble table top is ideal for this, or a sheet of formica, aluminium or mild steel. If the form is spherical, it is wise to place a small cylindrical spacer underneath as in figure *84A*.
3 Now make a wall around the form to retain the molten material. This can be a clay wall, linoleum or metal foil, or anything impervious and flexible. Seal any opening which might allow a leak, with clay or Plasticine.
4 Pour the moulding material over the form to fill the enclosed shape. The thickness of material should be a minimum $\frac{1}{2}$in (13mm) all over. Allow this to cool.
5 When the material has cooled remove the retaining wall, and peel the mould from the form. To do this and to be able to open it up satisfactorily for removal of the original and subsequent filling, it might be necessary to cut through the rubber-like mould on one side. The mould is now ready for use.

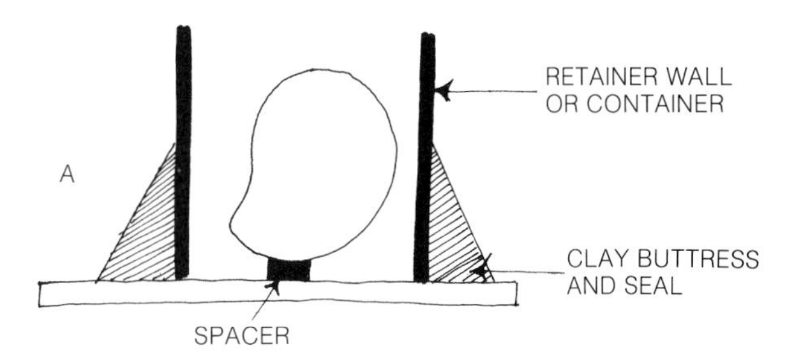

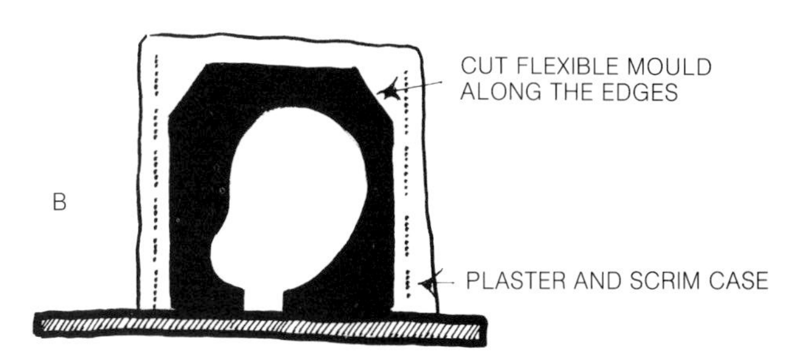

84 Simple PVC mould. A: the material is poured over the form held within a container. B: the simple plaster of Paris jacket placed around the cooled PVC mould.

Some sculptors prefer to put a plaster and

scrim jacket around the mould, after taking away the retaining wall (84B). This makes a case to hold the mould securely in position whilst it is being filled, and is particularly useful when working with a very soft compound, i.e., one with a low melting point.

This elementary process can be made complex when adapted to accommodate a wide variety of forms. The traditional method of making a case and mould, which has been developed for use with gelatine, is still, however, the most satisfactory technique. The method for making such a mould is at first glance complicated, a fact which often deters the student, but upon examination the processes and techniques involved prove to be very rational, sensible and orderly, as indeed are most of the basic skills in sculpture. The aim in making a flexible mould is to achieve an even distribution of material around the form, which must be retained in perfect register and shape by a plaster of Paris case or jacket. This jacket at the same time allows the flexible mould material to be easily withdrawn, so that it can be peeled in turn from the positive form. The mould and jacket must permit proper access to the surface of the mould to enable it to be satisfactorily filled, according to the casting medium. This description is the basic principle on which good and practical flexible moulds are made. The procedure for making a mould using polyvinyl chloride hot melt compounds, is as follows: (at relevant points any difference in the technique using the traditional gelatine, will be indicated; the process for both is similar)

1 Prepare the form to be moulded. If this is of a porous material, the surface must be sealed. This can be done by applying a solution of shellac and methylated spirit, followed by a fine film of oil, Vaseline or tallow. A patent sealer may be obtainable from the manufacturer of the PVC compound. Shellac and oil is quite adequate for gelatine. If porosity is not sealed, the mould will be full of tiny air bubbles, and virtually useless as far as good reproduction goes.
2 Place the original or master cast on a large board, on a modelling stand or banker. Prop the master cast in position with small lumps of clay. Decide where the dividing lines will be, and draw these on the form. Plan the mould and sections carefully, deciding the order and size of pieces, then build up to the first line with clay. Leave a small space between the clay and the form. This should result in a continuous band of clay about 3in (76mm) wide around the form. Make the surface of the clay band fairly smooth.
3 Looking down on the form determine any undercut shapes or textures and fill these with clay. The greasy surface will prevent the clay from sticking to the master cast (French chalk will also do this, and is all that need be used if the surface is not porous; oil is needed for gelatine).
4 Cover the original with an even layer of clay, approximately ½in (13mm) in thickness. Do this either by rolling clay into long strips and by carefully coiling, building up an even layer, or by rolling out a slab of clay and placing this carefully over the master surface. The reverse face of the clay thickness needs to be made smooth, and the form of this must slope inwards, to make the proper draw from the case possible.
5 Make conical shapes of clay and place them about on the clay thickness. Place a thick cone of clay, wide base down, at the highest point of the form. Then place thinner cones at the lower points. The number of the latter is determined by the area of the original; large forms require more, but a maximum of four will usually prove sufficient. The cones should rise about 1in to 2in (25mm to 51mm). These cones will eventually become the pouring space, or runners and risers, which control the introduction and proper

flow of the moulding compound.

6 Around the edge of the clay shape on the clay band, a lip of clay is built and lozenge shaped clay wedges placed (80C). These will provide the keys to give proper register to the piece of mould, locating one to another.

7 Cover this arrangement with an even layer of plaster of Paris, down to the clay band, making the topmost surface flat, and capable of standing. The shape of the plaster on the clay band is shown in figure 85B.

8 When the plaster has set and hardened, trim the case at the edges and top surface, to make a tidy job, then turn the whole work over to stand on the board. Remove the clay band and any loose clay. The plaster band now revealed must be trimmed to ensure a flat surface, locating keys cut into its surface and painted with a release agent.

9 The processes of building an even clay thickness, placing the clay cones, making the locating keys around the clay shape and the lip, must be repeated. Then the plaster jacket should be made, with a flat top, making sure of the shape of the plaster at what is now the seam between the case sections, which should be as shown in figure 85B permitting clamps to be placed to secure the case.

10 A third piece of mould must now be made without a clay thickness. This should have been planned when designing the mould. It is made in the form of a plate, which will register directly on the form or open end of a form. This provides proper register of the form and permits an opening in the mould giving access to the mould surface for filling. On a spherical form this is a straightforward process, but make sure that there is a release agent on the surfaces. On an open-ended form, which is hollow, this plate is made by first packing the hollow with paper then sealing around this with clay. This prevents plaster and moulding compound from getting inside the master. The plate is made to register on the presented edges of the open form, which should be treated with a parting agent. This provides the accurate register of the master in the case, and is useful for remaking the mould. It will be seen from figure 85A that this plate also provides a controlled stop to the flow of molten material.

The case is now complete. Next remove the clay from between the plaster and the original, replacing this with flexible material.

11 Remove the plate and one side of the plaster case. The clay thickness should pull away from the original, with the plaster, leaving only the clay plugging the undercut areas (if any).

12 Remove the clay thickness from the plaster case. Clean the case taking off any blemishes. Trim the tapered holes through the case, which will have been made by removing the clay cones. Widen if necessary the pouring hole. For gelatine, the jacket needs to be sealed with a solution of shellac and methylated spirit, then oiled.

13 The actual mould seam edge now has to be cleaned up and precisely made. This is done by modelling the clay in the case that remains around the master cast. Remember that where the clay touches the master cast, the precise division of the mould and the eventual seam flash on the cast is determined. Consequently as fine a division as possible should be made by a crisp edge where the clay and master meet. Into this clay seam registration aids must be made to locate the mould pieces accurately. Studs or a continuous ridge are made to make the register (85C). These are of course made as negative shapes in the clay.

14 The surface of the original should be wiped and cleaned. No oil need remain when PVC is used as the moulding material. For

gelatine, however, there needs to be a very fine film of oil on the surface.

15 When the seam has been cleaned and the locating features made, the prepared plaster case can be replaced and clamped in position. If there is an end plate, this is replaced and clamped also.

16 Over the pouring and rising holes in the plaster jacket, make a series of funnels. These can be made of clay to fit each job. Professional mould makers use prefabricated funnels, made to pre-determined set of sizes. These speed up that part of the process. The funnels are secured and sealed with clay.

17 The molten moulding compound is poured into the space between the case and original by means of the largest funnel placed over the pouring hole at the highest point of the form. Continue pouring until all the funnels are filled. The pour should be gentle and constant, unhurried and directed into one spot in the pouring funnel. When filled, allow the mould to cool and solidify.

Gelatine is ready for pouring when it has cooled to blood heat. This can be tested by feeling the bottom of the bucket with the inside of the wrist. Remove any scum from the surface before pouring.

18 When the first part of the mould has cooled repeat the whole process from 11 to 17 on each remaining piece of mould, until the flexible mould is complete. No separator is necessary between the seam faces of PVC.

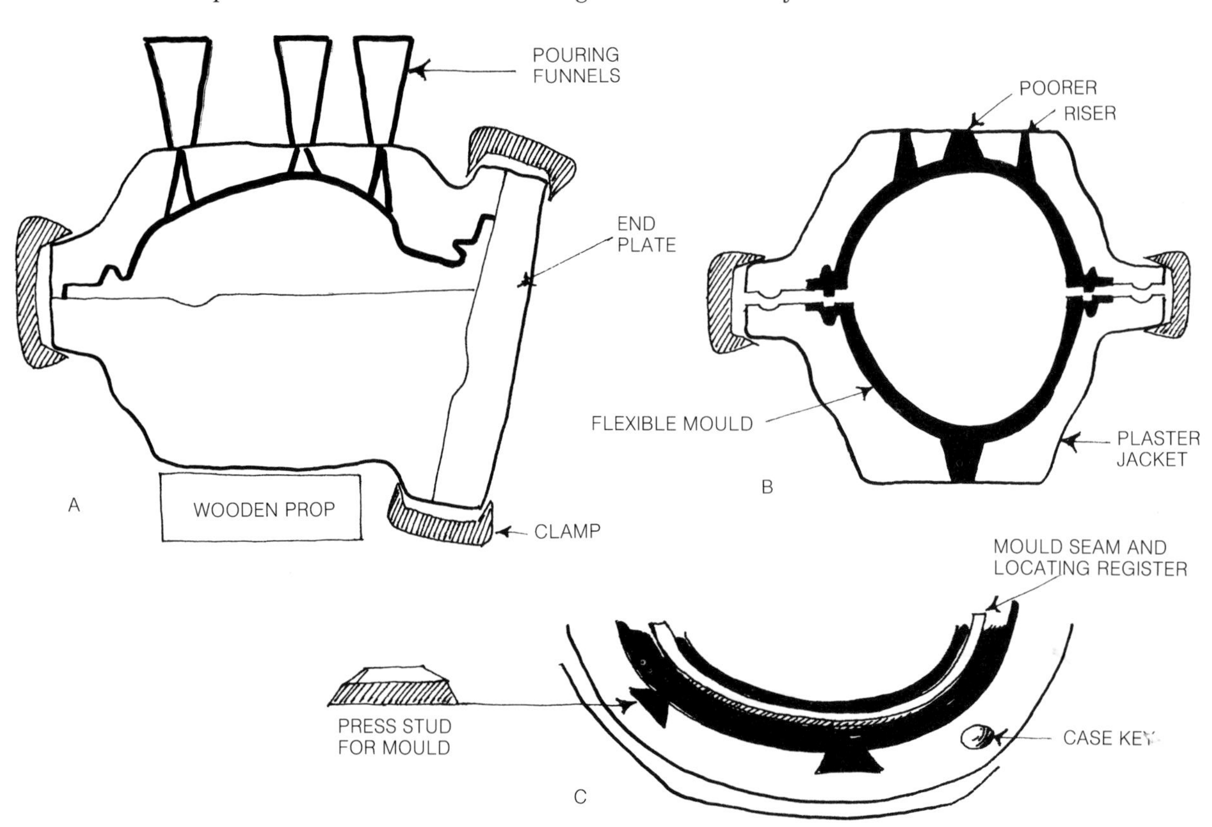

85 PVC or gelatine mould. A: the thickness of clay over the original. Pouring and rising holes and funnels placed over them to pour the moulding media. B: the complete mould thickness and locating keys to mould and jacket. C: details of these systems, showing the mould jacket clayed and secured with joiner's dog clamps.

To be certain of good separation dust the seam with French chalk before pouring its neighbour. For gelatine, the surface between pouring must be separated by the use of oil or tallow.

Points to remember:

1 A well-made and well-designed case is essential.
2 The thickness of clay should be adequate, not too thick or thin, and tapered to be able to withdraw it from the case.
3 The pieces of case must register properly and the flexible mould must fit securely into the case. The lip and lozenge shapes should ensure this.
4 The seam of the mould must be precise and must register properly; any discrepancy at this point would result in poor volumetric reproduction.
5 The pouring and rising vents through the plaster jacket should taper, and the funnels placed over them should be of equal height.
6 The molten media must be poured steadily.
7 Proceed with caution and plan each step thoroughly until the work is done.

When the pouring of each piece has been carried out and the mould completed and cooled, remove each piece, case and mould and take out the original. Trim off the surplus material, for instance, the excess at the pouring cone; these can be trimmed to leave short stubs only. The mould is now ready to be used. No release agent is necessary between hot melt polyvinyl compounds and casting media.

Gelatine, after the removal of the original, needs to be treated to get the best results. First a dusting of French chalk between the case and gelatine is required to allow easy removal of one from the other, and to prevent deterioration of the gelatine, due to moisture. The actual mould surface can be treated by first applying a solution of alum in water, to harden the surface and then by applying a thin dilute clear cellulose, to protect it further from moisture in the filler. Lastly an application of oil is made to give a final release from plaster; if from wax, a dusting of French chalk will suffice.

Polyvinyl chloride moulds, if treated with care, permit an almost endless number of castings to be made with various fillers. Gelatine permits very few, in fact, only approximately four satisfactory plasters or waxes can be made from a single mould before it needs to be repoured to make a fresh one. Some very careful moulders can just manage an edition of six waxes from a single gelatine, but this is not common and should not be expected. Once the case is made, however, it is a relatively simple matter to pour and produce subsequent moulds from the master cast.

When the flexible material, of either kind, is cut up and used again, the plaster of Paris case can be retained. Providing that storage space is available the case can be stored and used

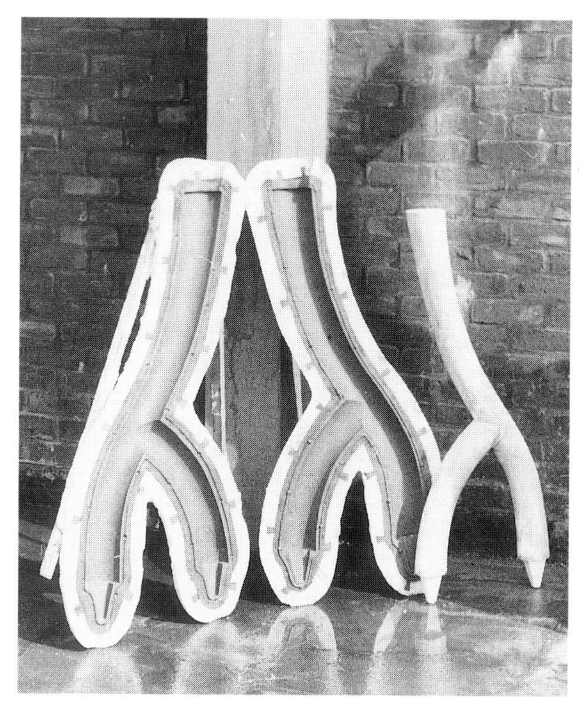

86 PVC mould and case, with the master cast from which the mould was made. The registration keys can be clearly seen, and the press stud location points in the plaster cast. The middle section of the sculpture has been cast from the flexible mould.

87 The completed sculpture in polyester resin and glass fibre, by Richard Rome, Chelsea School of Art.

again, to produce another mould from the same original or from another work of the same dimensions. With gelatine it is important to be able to make more than one mould to complete an edition. To do this the original is simply placed in the large piece of the mould case, on top of clay spacers which prevent it from touching the case. The plate of the mould can be used to help locate the master cast in the mould case, because it is in fact an impression of the surface or edges of the master, and seam sections of the case. The seam of the actual mould is made good around the jacket and the original, by filling with clay and modelling a precise edge. Remember to make the necessary allowance of locating studs or ridge. The remaining parts of the plaster jacket are next clamped firmly in position. Pouring funnels are placed and sealed and the material poured. When cool and firm, the clay can be removed and that part completed by simply repeating the process. In this way a mould can be made in an existing case. This accounts in part for the collection of plaster jackets to be seen around the plaster shop of most foundries, using the cire perdue technique for bronze casting.

Blemishes on the surface of a PVC mould can be trimmed and made good. This is another advantage of this material. A hot spatula can safely be used to touch up, remove a blemish, or attend to an offending surface on the mould. If it is found that an undercut, taken in by the moulding material, is too great, causing too great a strain on the casting when removing the mould, the mould can be cut. With a sharp knife, a cut can be made cleanly through the mould. This will make it possible to remove the undercut sections of the mould piece, thus avoiding strain on the casting. It will be found that the cut pieces, providing no dust gets between them, will stick and hold firmly together sufficiently well in the mould whilst this is being filled. Many difficulties can be overcome in this way.

Room Temperature Curing Compounds

Among the newer materials developed since World War II for specific industrial needs have been a number that have proved useful to sculptors as moulding and casting compounds. Silicones, silastomers, polysulphides and polyurethanes are now identified under the generic heading of Room Temperature Curing or Vulcanising compounds (RTC and RTVs). These are synthetic materials as distinct from latex, which is an air drying natural rubber. They were developed to fulfill particular job specifications or replace previously cumbersome materials, or simply to increase the range of products available to industry. Their common industrial application was either insulation of various kinds or protection against vibration and adverse weather conditions. These substances can be poured, painted or trowelled, according to the mixture used, and each will be firm but elastic when cured in the final set form. This means they are water repellant and elastic. These qualities are entirely suited to mould making. Industrially they are usually applied by dipping the item to be treated, into a preparation of the material. Alternatively the prepared cold curing compound can be poured over the item that is contained in a particular vessel or mould.

It can be seen that these industrial applications are sympathetic to sculptural needs. To the techniques of dipping and pouring add the further possible processes of painting and building up, and a moulding material with a wide potential is seen. Moulds made from such materials are relatively expensive, since the basic raw product is not cheap, and moulds cannot be broken up to be used again when finished with. The mould when properly made and carefully stored, however, will last indefinitely, and a great number of castings can be made from a single mould, without damage to the surface. Provided the mould is made with precision and care, certain advantages are gained over other moulding media; a high degree of flexibility for instance, plus very high quality reproduction of surface detail.

It is difficult to give particulars of specific materials, because of their wide range and because they are undergoing change and development in industry all the time. It is, after all, the industrial demand that governs the manufacture and development of specific materials. A characteristic of industrial development today is that fact that substances and processes become rapidly obsolete due to rapid development. Enquiries should be made from manufacturers for information regarding such materials and innovations, whether the substance used has been liquid or paste, cured by exposure to the air, as in latex, or a liquid or paste cured by the addition of a catalyst, as in silicone rubbers and silastomers. The moulding methods developed have been a combination of some of the traditional processes and techniques.

The original should be non-porous, to gain the best results. Porous forms will cause the mould surface to become aerated and spongy, and of poor quality. Porous items should therefore be sealed before applying the moulding substance by using shellac solution, liquid wax, tallow or a proprietory sealer – many are now available, so seek advice from the supplier or manufacturer of the chosen material.

Pouring the moulding compound to encapsulate the original, held in a suitable flask or vessel, is an obvious technique requiring little further explanation. The flask can be retained to maintain the form without volumetric distortion, which might occur once the support is removed. A case of plaster can quickly be made around the rubber. Cases can be made from a laminate of polyester and glass fibre; this may be suited to the permanence of the mould but is possibly an unnecessary expense. Once the mould has cured it can be cut, with care, using a sharp, thin-bladed knife, to release the original. Stability is important to a flexible mould enabling it to maintain the correct volume and balance of particular form, so it is important to make the means of support before cutting the mould.

Dipping the original into a synthetic rubber compound achieves a skin of material over the surface. The dipping should be repeated more than once, and if possible the original should be agitated whilst immersed, to prevent air bubbles forming on the surface. The film of flexible material on the master must be reinforced so that it is rigid but keeps its flexibility. This is done by adding substances to the liquid to make it still. With latex rubber and silastomers, this means adding fine sawdust or some other inert filler to make a consistency that permits the build up. Silicones, silastomers, polysulphides and polyurethanes are all manufactured as both liquids and pastes, thus enabling a fine detailed surface to be dipped or painted using the material as a liquid (with a fast curing catalyst) followed by a build up using the substance as a paste. Fibrous material can be included in the build up to give added strength to the mould, but most of these synthetics have a high tear factor and so may not need much reinforcing. Deep, fine, undercut forms will need strengthening and cotton gauze or scrim is a useful additive in such cases. Metal is not often needed in building up synthetic rubber moulds, but if it is used, care must be taken to see that it is placed so as not to hinder the cutting of the mould sections.

A moulding technique devised by John Pappas in Michigan employs the painted surface and a paste build up as described, but a refinement he uses is to cast strips of the polysulphide. The strips are made in various widths so that he can use them according to the size of the piece being moulded. These strips are included in the final layer of the build up and straddle the eventual mould split line. The plaster jacket is made using a clay wall placed in the middle of the raised strip. When the case is complete and hardened the sections are removed and the rubber compound cut along the same division. When this is removed from the original work, it can be placed in its appropriate cast section, which, because of the precast strip on the seam edge, will provide the perfect register, retaining the flexible mould securely.

As already indicated, it may be possible, according to the shape and nature of the original, simply to dip and paint the synthetic rubber to the surface of the master cast, and in this way build up a thickness and a shape that has no undercuts. If a compressed air line is available, it will help to blow the painted first coat, forcing it into all the detail on the surface of the sculpture. If this is possible, then a simple layer of plaster of Paris can be made over this to provide a case.

A kind of piece mould technique can be used to make moulds with these synthetic rubbers. This is done by deciding upon the pieces to be made, remembering that the material permits undercut forms to be included in a section of mould are made as in figure 82 and suitably determine the split line between pieces, make each section separately, painting, building up and reinforcing. Remove the Plasticine or clay wall when the materials have cured, and apply a parting agent to the seam surface; oil, Vaseline, tallow, etc. Place more clay or Plasticine walls if necessary, and make the adjacent sections of mould. Carry on in this way until the mould is complete. If the sections of the mould are made as in figure 82 and suitably reinforced, there will be sufficient rigidity. It is necessary when making a mould in this way to make few pieces, being in mind the flexibility of the moulding media. This can be turned completely inside out and has fine elasticity. Deep undercuts and heavily textured surfaces therefore can be moulded simply.

Because of the comparative newness of the material, this description is inevitably brief. Experienced mould makers, however, will have little difficulty in adapting known techniques to these synthetics and relating to them their particular needs. Because of their relatively high cost these substances are used with care; they are very useful, however, when a very delicate work must be moulded, and more traditional processes are either not available or unsuitable. An instance of this is the moulding from very small delicate wax originals. There are no other materials that permit a flexible mould to be made easily from a wax master, and this could prove to be a serious deterrent to working directly in wax. Another instance of the extreme suitability of a cold curing rubber compound is in the archeological field. It is possible with these materials to make a flexible mound *in situ*; a fact that has proved invaluable to archeological recording and a further indication of the invaluable quality of the synthetic rubbers. Specific information regarding formula and technique should be sought from the suppliers of the materials. I have mentioned the extreme diversity of the materials produced industrially and it is wise to find out what characteristics a particular product may have.

Expanded Polystyrene

This is a foamed plastic made by extruding molten polystyrene containing a blowing or foaming agent, methyl chloride. Heat causes molecular expansion of the blowing agent and when set the material is 40 times greater in volume than the starting substance. It may be made with an enormous proportion of its expanded volume consisting of air, up to 95 per cent with only one-tenth of the density of balsa wood or cork. The manufacturers and suppliers can deliver this in an expanded block form, cut to almost any size required, in expandable beads to be moulded or even preformed to specific requirements.

The application of this material in sculpture is varied and interesting, and has an extremely large potential. It can be cut by means of a heated resistance wire (*88*), and it can be modelled to some degree by using heated metal spatulas or electric soldering irons. When using the latter I have found it necessary to cover the iron with a brass or copper foil and copper wire (*90*). This protects the heating element and makes for a very useful tool. A sharp domestic cheese grater, the type held in the hand on which cheese is rubbed, is also a very useful tool. Plaster raspers, providing they are new and sharp, can also be used to good effect. A very sharp knife is invaluable.

A word of warning when using any of the foamed plastics; because of the lightness and resistance to acids they present certain dangers to health. Dust and fine particles tend to float, and consequently are breathed into the lungs. Their acid-resistant properties render them impervious to body acids also, so when working with these foamed plastics it is advisable to wear a mask over the mouth and nose. These can readily be obtained from industrial suppliers.

Expanded polystyrene is a versatile substance. It can be used as a foundry pattern, as a former and as a mould. It is useful in producing definite kinds of form and texture. In the building industry it can be used to elaborate the casting faces of shuttering, making it possible to produce a patterned or sculptured relief directly on site. Panels of this material can be fitted to the inside surfaces of a shuttering, the panels being in fact mould surfaces, the design cut and modelled in reverse into the panels. The designs can be made *in situ*, or elsewhere and brought to the site, and fixed to the shuttering. When this is complete the wall or column, or whatever architectural feature it is, can be made by filling with concrete, placing the reinforcement and pouring the concrete (perhaps with a little more care than usual, in view of the patterned mould surfaces). The concrete is vibrated to ensure that the fines, the smallest particles of the concrete mix, are cast against the mould

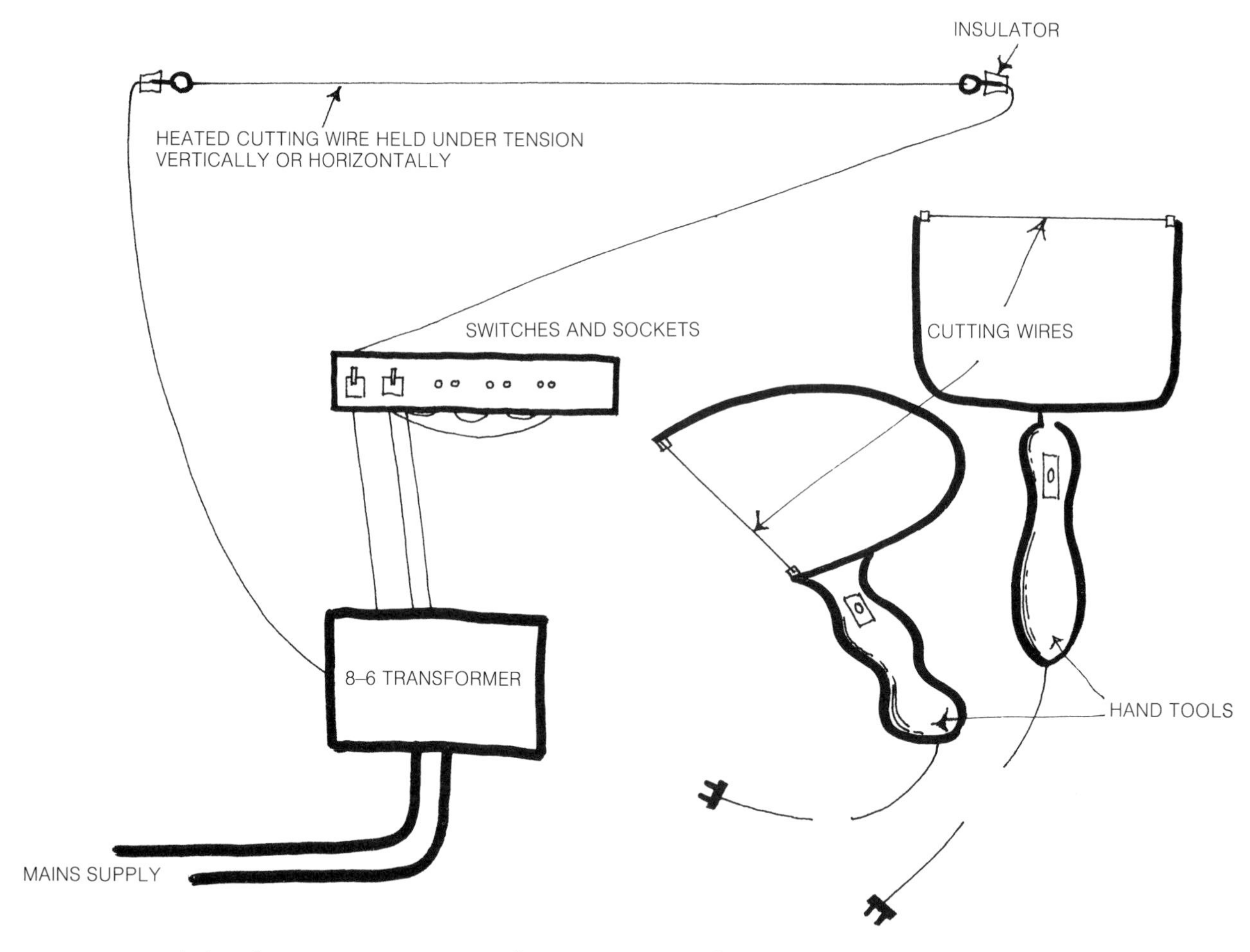

88 Expanded polyester cutter: various hot wire cutters designed for use with expanded polystyrene or rigid polyurethane foam.

surfaces, and to eliminate air bubbles. When the concrete has hardened and cured, the shuttering is removed, together with the polystyrene. Any that might remain on the concrete surface can be cleaned off with a blow lamp or styrene. It volatalises under heat and disappears or it is dissolved by the styrene or cellulose thinners.

Designing negative form using expanded polystyrene can be extremely interesting. It can be built up in blocks, modelled and cut; pieces can be glued together, using a special polystyrene adhesive as most others are antipathetic and will dissolve the polystyrene; pieces can also be added by simple pinning. It is possible to cast plaster of Paris, cements and concrete, and various synthetic resins from expanded polystyrene moulds.

No traditional methods exist for producing a mould from this material because of its very youth, if not infancy. Experiment will produce a great variety of techniques but as a mould it has a particular application for use in producing sculpture for an architectural environment. This can be adapted, refined and given a greater application. Its use as a former and as a foundry pattern maker is, at this moment, I think the most interesting.

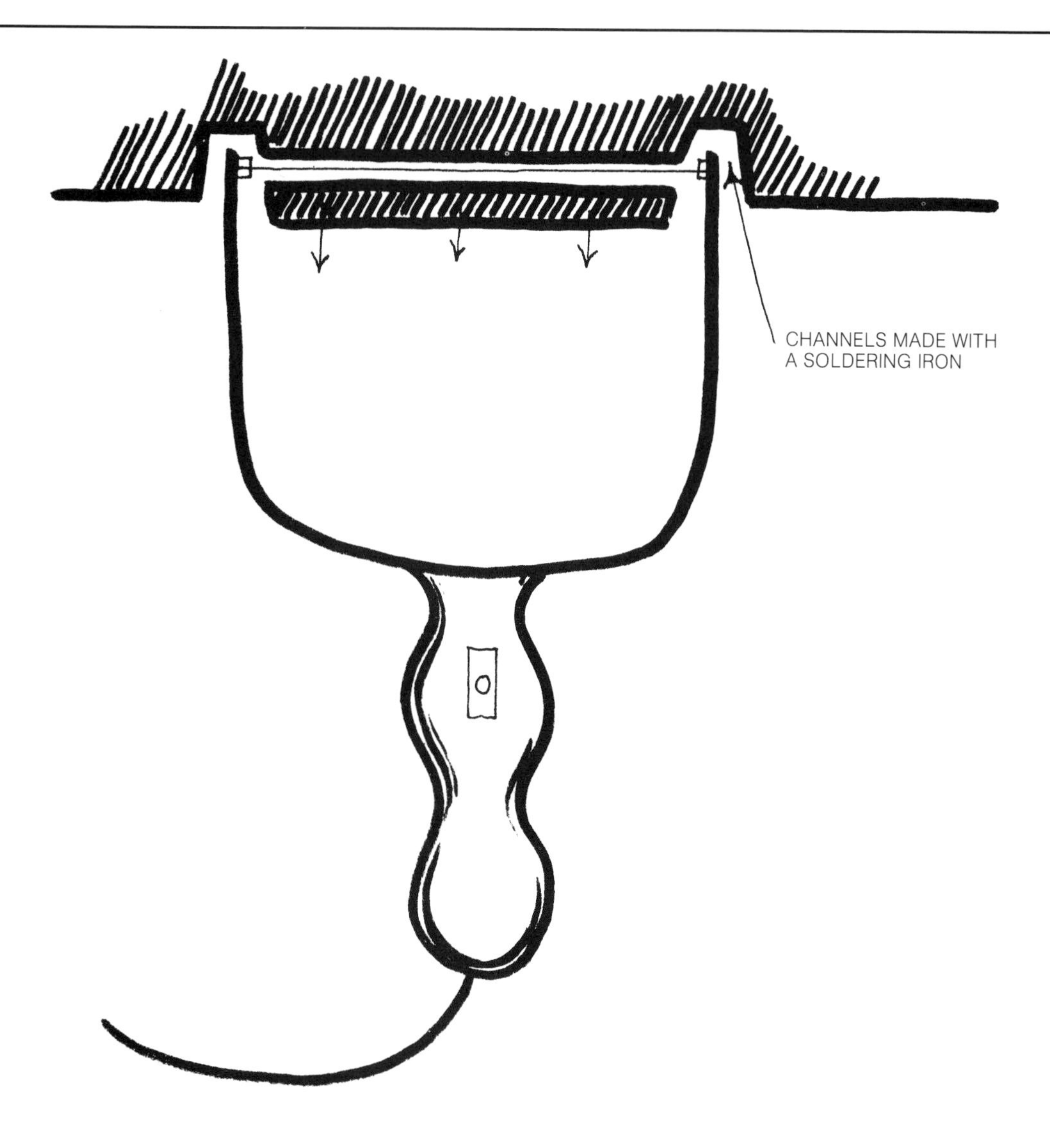

89 Hot wire cutting. A channel can be made with a soldering iron to permit access of the hot wire. In this way the foamed polymer can be cut and shaped.

A former assists the manufacture of a final form. The basic shape can be made of expanded polystyrene, taken to within $\frac{1}{2}$in to 1in (13mm to 26mm) of the final surface; the final form and surface texture being modelled over this basic shape or former (93), which may be of plaster of Paris, cement, resin or wax. The plaster of Paris and the wax are dealt with here as they are, generally, intermediary stages in the metal casting process. Cement and resin are modelled direct, to result in a final form.

The following description of a work of mine

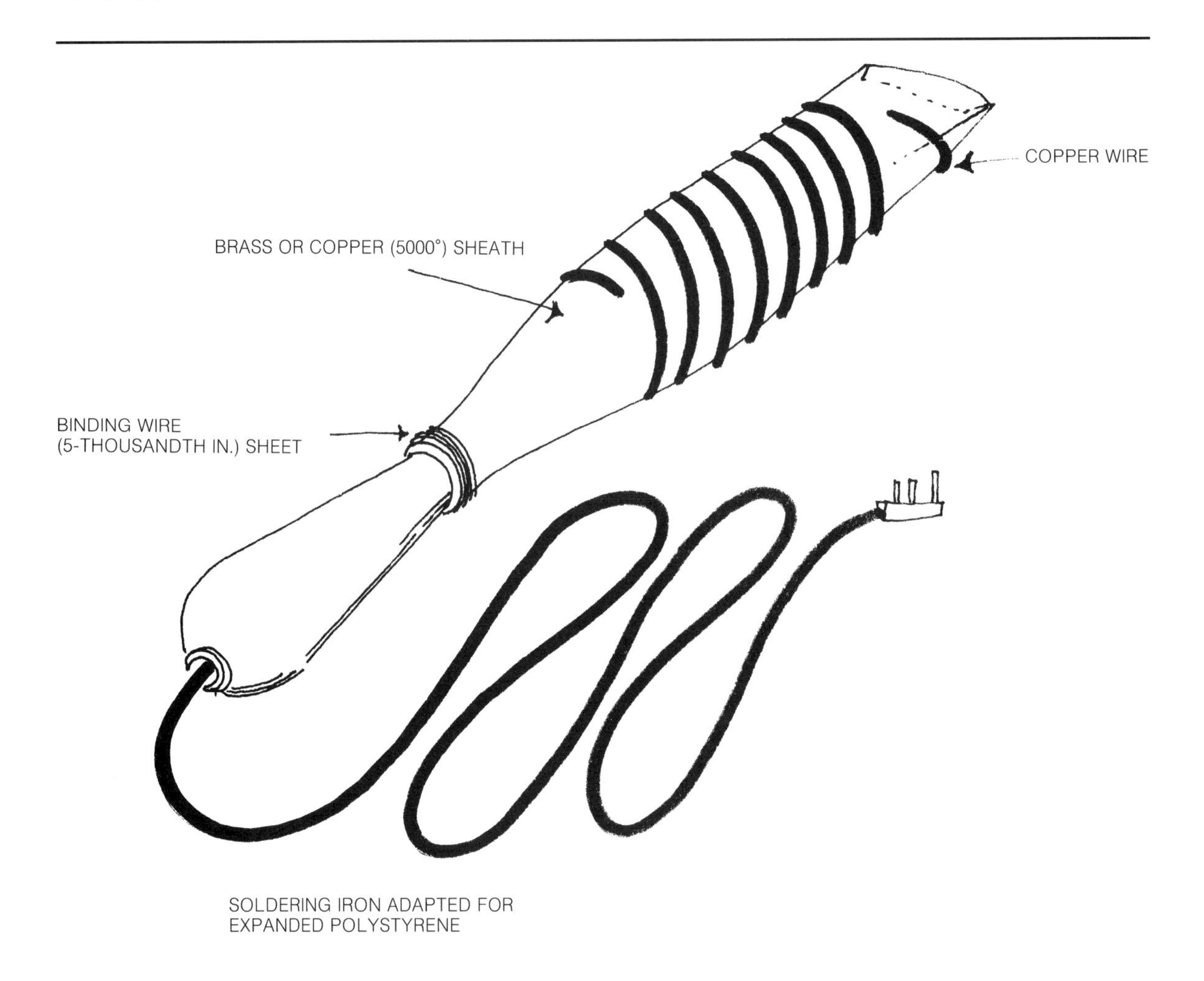

90 An electric soldering iron adapted for use on expanded polystyrene.

will illustrate this use of expanded polystyrene. The problem was to design a relief sculpture of aluminium to cover a large wall area, the size of the work being 18ft by 14ft (5.5m by 4.3m), with a high projection. It was to be produced at a fairly low cost. I did not want to model the original in clay and have to waste mould and cast the pieces for the foundry. I wanted to work the forms and surfaces directly in plaster, then to have them cast in metal, via the sand moulding process of an industrial foundry, rather than an art bronze foundry, and after casting continue working on the metal. I could model direct in plaster over a metal armature, but the time involved in making a precise armature would be too great. The sand moulding flasks could take only panels of 4ft (1.2m), therefore subdivisions would have to be allowed for. I therefore bought blocks of expanded polystyrene, varying in thickness, cut to 4ft by 4ft (1.2m by 1.2m) and smaller to fit my design. From this material, supported against a wall, I modelled the relief. I took this to within approximately $\frac{3}{4}$in (19mm) of the final surface. This final thickness and surface was modelled direct in plaster of Paris,

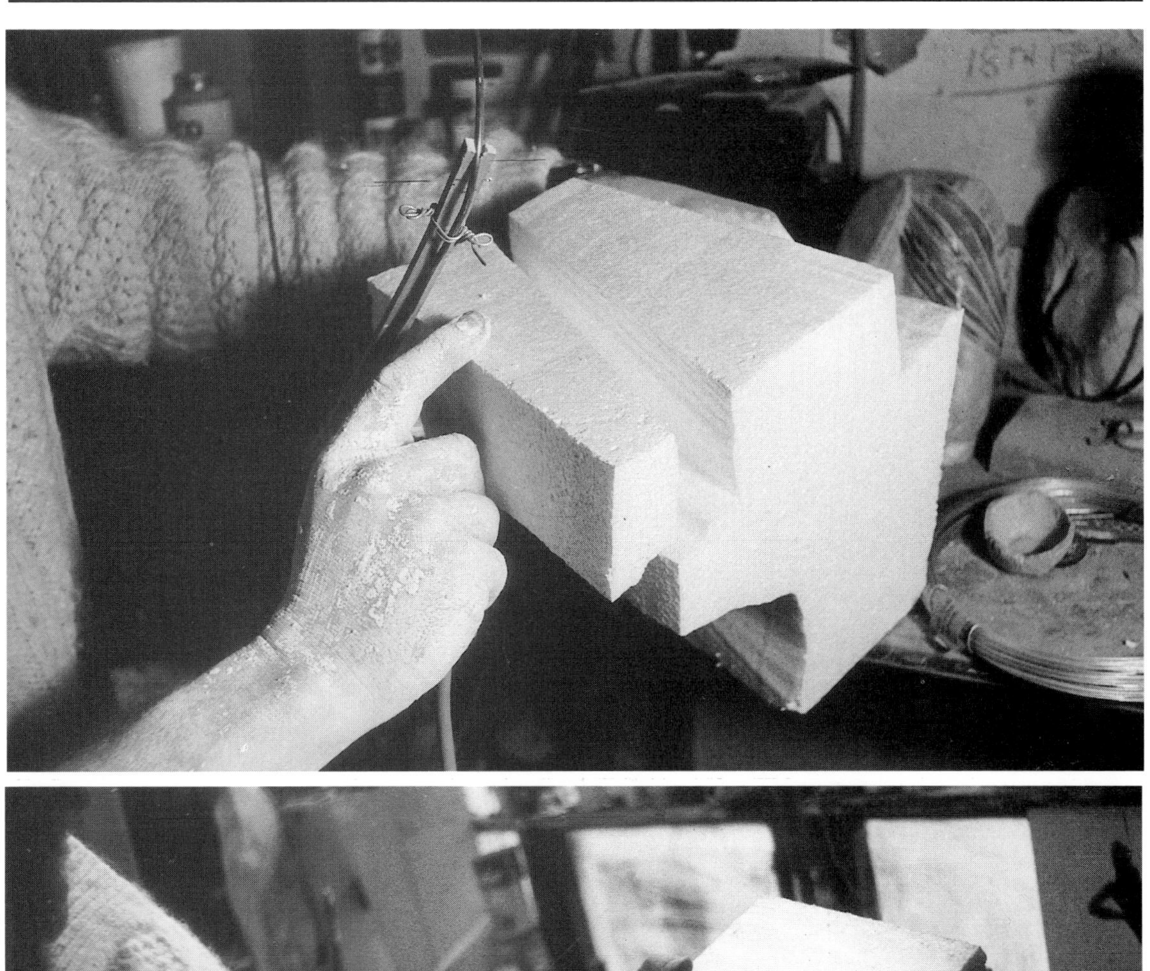

91 and **92** Cutting expanded polystyrene with a hot wire.

93 Working the large relief with a sheathed electric soldering iron on expanded polystyrene.

reinforced with mild steel rod and jute scrim. Detail, surface texture, all were modelled at this point. After modelling the panels in the polystyrene stage, each was removed from the frame, and the plaster modelled, providing at the same time register at the corners of each panel. Adjacent surfaces were cast against their partner to gain good registration and matching surfaces. When all panels were completed the relief was re-assembled and the final surface texture and detail, mass and finer forms, modelled in plaster to make a unified whole.

What I had achieved at this point was a plaster of Paris original or master, modelled over a core, or former of expanded polystyrene; this original being already sub-divided for the foundry. The polystyrene could be removed from the back of each plaster section, leaving what constituted the foundry pattern. Whilst the polystyrene remained it provided extra support to the pattern, useful for making safer

94 Scraping to form the expanded polystyrene. This useful tool is a stainless steel cheese grater.

95 Making the mild steel reinforcing to strengthen the plaster thickness, seen in the adjacent section.

96 The expanded polystyrene revealed on the back of the plaster thickness. This is then removed to enable the sand mould of the reverse face to be made. (Shown here in the foundry.)

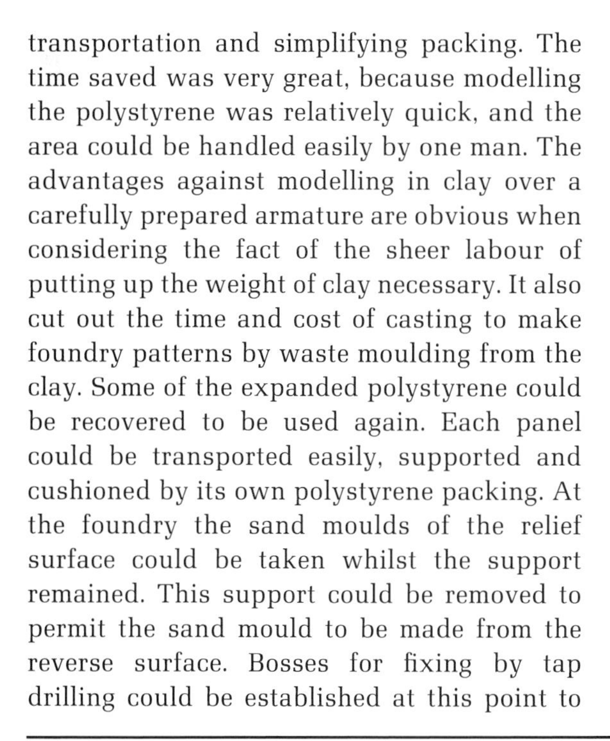

transportation and simplifying packing. The time saved was very great, because modelling the polystyrene was relatively quick, and the area could be handled easily by one man. The advantages against modelling in clay over a carefully prepared armature are obvious when considering the fact of the sheer labour of putting up the weight of clay necessary. It also cut out the time and cost of casting to make foundry patterns by waste moulding from the clay. Some of the expanded polystyrene could be recovered to be used again. Each panel could be transported easily, supported and cushioned by its own polystyrene packing. At the foundry the sand moulds of the relief surface could be taken whilst the support remained. This support could be removed to permit the sand mould to be made from the reverse surface. Bosses for fixing by tap drilling could be established at this point to facilitate assembly and securing on site.

By using this method I was able to produce a sculpture relief, large in area and at a reasonable cost (*91* to *96*).

Refinements of this technique will obviously present themselves. Plaster modelled over an expanded polystyrene former offers unique opportunity for making patterns suitable for sand moulding. The plaster can be modelled to a completion, then simply sawn up to make pattern pieces, removing the polystyrene from each piece. If it is not necessary to cut through the original, the volume and shape being such as to permit a hollow casting to be made in one piece by placing a simple core, the polystyrene can be removed by dissolving or heating.

Wax can be used as a modelling medium over expanded polystyrene and in this way a wax original can be made, using a technique similar to that with a pre-determined core

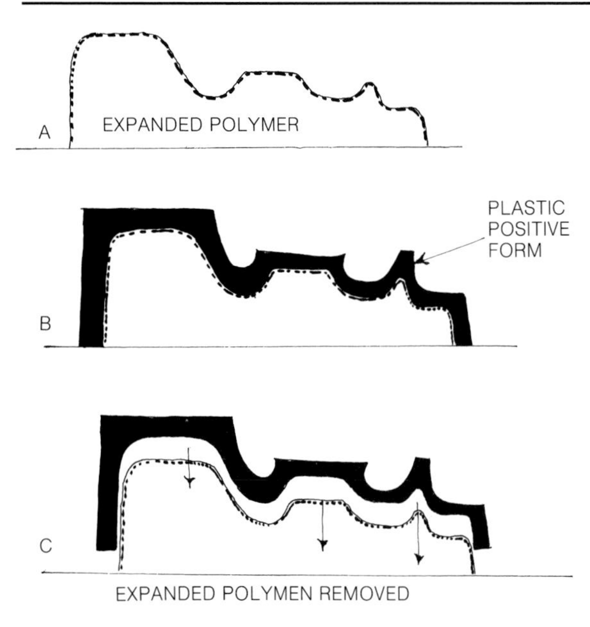

97 Foamed plastic former. A: the plastic is moulded to make the former to within $\frac{1}{4}$in–$\frac{3}{4}$in (6mm–18mm). B: the positive form is built up over this and finalized. The material can be plaster of Paris, cement, wax or polyester resin. C: the former is removed leaving the pattern or hollow positive. (The former can be retained within a final form.)

process; the advantage in this case being that polystyrene can be removed by dissolving to be replaced by a conventional core. Whilst the wax is being modelled the whole shape can be changed, thereby offering greater freedom than with a rigid, unchangeable former.

No doubt sculptors will evolve personal methods for using expanded polystyrene. It is relatively new to industry and newer to sculpture, presenting a great potential as sculptural media. The following procedure, which I use to produce plaster patterns, can also be used to make positive forms in resins or concrete:

1 Carefully plan the work in hand, bearing in mind that the substance can be easily cut to pre-determined sizes by the supplier. Order the material and prepare the planned sections.
2 Model the former using the tools best suited to it. Heated resistance wires, soldering irons, heated metal spatulas, raspers, graters and sharp knives. Take the former to within $\frac{1}{2}$in to $\frac{3}{4}$in (13mm to 19mm) of the final surface. Detail is relatively unimportant at this stage.
3 Model to the final surface, form and detail, in whatever material is selected. Be careful not to build too great a thickness; keep to approximately $\frac{1}{2}$in to $\frac{3}{4}$in (13mm to 19mm).
4 When this final form is settled and hardened, remove the former. This can either be cut out, removed by heating, or by dissolving with a solvent. (See lost pattern casting using expanded polystyrene pages *148* to *9*).

Rigid Polyurethane Foam

This material is similar in many ways to expanded polystyrene but has a finer cellular structure. This means that it will permit finer forms to be made. It can be used in the same way as expanded polystyrene, in all the applications I have mentioned, with the advantage of greater finesse.

The sculptor, John McCarthy, has developed a technique to produce formal designs using rigid polyurethane foam as a negative form; items are pressed into the foam to form a design, and polyester resin cast is then taken of it. The potential of this material, like so many of the plastics, is not yet fully realised, and it is certainly a material worthy of consideration and experiment.

IV Clay, Wax and Sand

The three materials clay, wax and sand are all suitable for use as moulding materials and are readily available. They are often overlooked, I feel, because of their familiarity. Among the first moulds ever made to produce positive forms by casting were moulds of clay. These were sometimes fired, becoming terracotta, and making strong durable moulds which were used to produce a kind of production line. Because of its strength and durability the terracotta mould permitted a large number of castings to be made. Such castings were of wax or clay. This was the basis of the technique of cast metal working practised by the Bronze Age artisans and by subsequent civilizations.

Each of the three materials dealt with here can be made soft and pliable and be used to take an impression. This impression will be accurate and held long enough to make a cast from it. Plasticine, which is a synthetic clay, is often used by over zealous individuals to make pirate impressions from relief sculptures. This is a social abuse of the material. Plasticine held in the hand, becomes soft and pliable, and can be pressed firmly against the relief to make an accurate impression. This subsequently serves as a mould from which a casting can be made. I have known such impressions, or moulds, taken right under the noses of museum custodians. These were taken for educational reasons by sculpture students wanting direct reference, but I do not advocate these illegal practices.

Clay Many sculptors today use techniques which are similar in principle and character to the pirate practice. Henry Moore employed this method to make moulds quickly from organic forms such as stones and bones, and by making plaster castings from the clay moulds he could proceed to make an image derived from that form, but which is purely the source of the idea. In this way Henry Moore could explore the many possibilities which may spring from a single source, using as his point of departure the actual organic form, and changing it by adding or subtracting.

This technique can be used to make castings from forms of almost any size and obviates the labour involved in modelling a facsimile of the form which is the basis of an image. This technique encourages a maximum freedom and at the same time a precise point of departure.

In my own sculpture I have found it useful to make clay moulds quickly from existing sculptures. In this way an alternative image or number of images can be exploited whilst retaining the original. Some sculptors require the confidence of having the first statement before making variants. This is possible by making a clay mould.

It will be understood that this is possible not only with relief sculptures but also with elaborate and complex forms in the round. The principle on which such elaborate clay moulds can be made is that of piece moulding (see Chapter II); every undercut form requiring a separate piece mould. The clay must of necessity be in good modelling condition, even in consistency, and free from lumps and foreign bodies. The same rule applies to synthetic

98 *Nude with Hands on Hips* by Degas, bronze. *The Tate Gallery*

99 *The Little Ballet Dancer aged Fourteen* by Degas, bronze. *The Tate Gallery*

clays, they too need to be of even consistency. The master cast from which the clay mould is to be made must first be dusted with French chalk or talc, the parting agent. The clay is then rolled or beaten out to make flat pieces for moulding. The beating, to make flat pieces, should be done on a surface dusted with French chalk. The pieces of clay should be fairly thick to allow the clay to be pressed well down and consolidated on the master surface. It is wise to make separate mould sections of undercut forms. Trim these carefully and dust them in turn with French chalk or talc. The remainder of the master cast can be moulded over these separate pieces making as many pieces of mould as necessary. The seams of the mould are made by carefully trimming the clay to fit well, one piece against another. Trim the seam edges with a knife and dust with French chalk. The back surfaces of the clay must be roughened to make a key. A reinforcement of scrim and plaster is then made on the back surface of each piece of mould (*100A*). A simple template is required next to register and retain in position the pieces of mould, ensuring reproduction of detail and volume (*100B*). When the template is removed the mould sections can be lifted off the master cast by means of the plaster backing; the suction between plaster and clay permits this. The seams of the clay mould when reassembled can be pushed together and made good.

Castings of plaster of Paris, concrete or wax can be taken from a clay mould. Polyester resins can be cast from clay too but the mould must be allowed to become leather hard to allow this; moisture being antipathetic to the gelation and cure of resins.

Taking the basic forms, practise making clay moulds, to be developed for personal use. The master cast must be of a hard material:

1 Dust the master or original with French chalk or talc.
2 Prepare the clay and make flat areas in readiness for the sections of mould.
3 Place the master on a base board or table top and support it.
4 Make the first piece of mould by pressing the clay sections firmly onto the surface of the master. Clay can be added to the section as it is being made, but it must be added to the reverse surface. The section is, in this way, spread and built up and creep marks on the mould surface are thus prevented. If there are any undercut forms, texture or detail, which may prevent lifting or draw of the mould from the original, these must be moulded separately. Trim the section of clay at the edges to make the seam faces. Dust these faces with parting agent, and key the reverse surface.
5 Reinforce the clay sections with scrim and plaster. Trim the reverse of this plaster to give a suitable surface for the template.
6 Turn the whole work over and make as many pieces as necessary to complete the mould, trimming each seam face, dusting on the parting agent, and placing the reinforcement to each section of mould. Finally make a

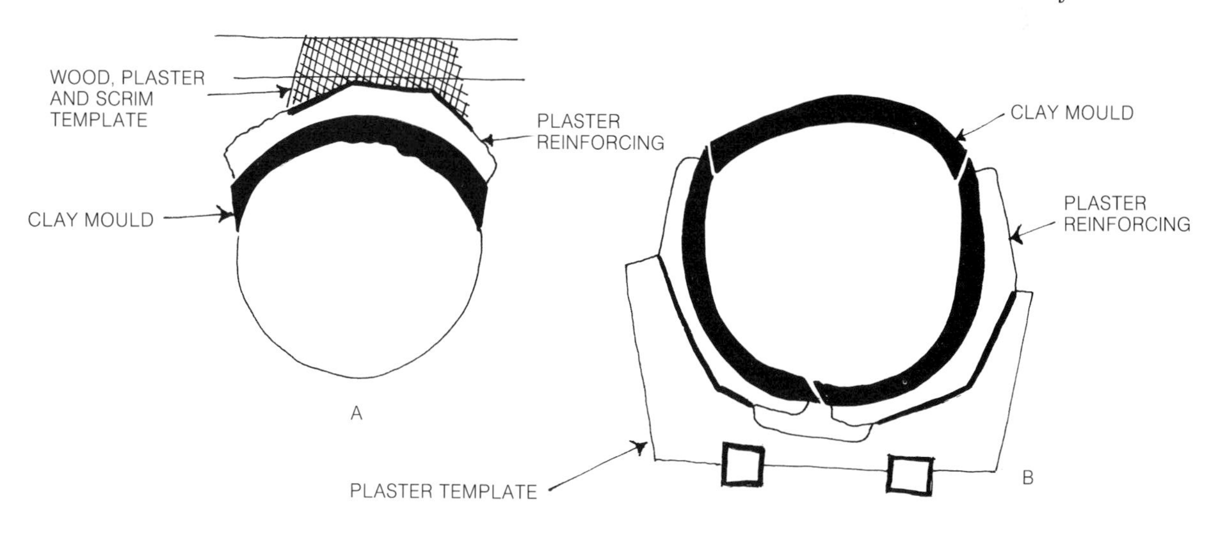

100 Clay mould. A: the clay thickness, plaster reinforcing and wood, scrim and plaster template. B: plaster template to contain mould sections.

suitable surface for the template to register on.

7 The template is made by taking impressions of the reverse surfaces of plaster backing with plaster of Paris. A release agent, suitable for the release of plaster from plaster, must be applied first. These impressions are connected by mild steel rods or wooden battens, fixed with scrim and plaster, in such a way as to form a definite cradle. This cradle holds the sections of mould in place.

The template described here is for a fairly large mould; simpler, smaller moulds require the simplest template register.

8 When the template is complete and hard, remove it. Remove the pieces of mould carefully from the master and fit them into the template cradle. Undercut pieces should be carefully removed and placed in position in the mould.

9 Make good any blemishes on the mould surface, which is then ready for filling. Access to the mould for filling should be designed according to the substance to be used.

Points to watch:

1 Clay of poor consistency.
2 French chalk accidentally getting between clay applications which need to adhere.
3 Adequate dusting of parting agent on the master surface and between seam faces.
4 Design of the scrim and plaster backing to cover just up to the edges of the clay section, leaving enough of these exposed to make a controllable seam (*100*).
5 Design of the template to make a satisfactory cradle.

Do not expect to make a perfect clay mould the first time; it does require practice.

Wax This is a very useful material indeed and can be used to produce high fidelity castings quickly; speed is always an important factor.

The wax that provides the most variety is micro-crystalline wax. This is a by-product from the refining of oil. The material is made ready for use by melting in a suitable vessel, i.e. a saucepan or bucket over a heat source. When it is molten a dye can be added to the pale honey coloured wax to make it a dense colour, so that the form may be seen or the thickness of application assessed. The wax can also be softened by adding heavy oil or petroleum jelly to make an easily modelled substance. This may be necessary to make good a fault in a mould surface or to help seal a mould.

The simplest wax mould is made by dipping a clay or synthetic clay original into molten wax a number of times, allowing the wax to cool between each dipping. When this has been done sufficiently to build up a firm thickness, $\frac{3}{16}$in to $\frac{1}{4}$in (4mm to 6mm), cool and harden the wax by immersing it in cold water. Cut through the mould to make divisions as necessary to remove the clay and provide access for the filling. Immerse the cut mould in water again to ease the mould open. Remove the clay and clean the mould surface. The mould can be sealed together by applying molten wax at the seams when the pieces are placed together. The mould can then be filled with a suitable material. When the filling has hardened the wax can be peeled from the cast; it may be necessary to heat it gently in hot water.

If the mould is to be filled by packing and not by pouring, the mould is put together at the same point in the procedure and secured with molten wax, using a brush and a heated spatula.

To make a wax mould from original forms other than clay, a dilute clay wash must be first applied. The wax should be built up whilst this is still wet. To make a wax mould stronger, reinforcing can be added. This can be of galvanised wire, wood and scrim. I will give a general procedure for making a wax mould, which may be practised on the basic forms. Small items can obviously be dealt with by dipping, larger items can be moulded as follows:

1 Prepare the original. If it is not clay or synthetic clay, apply a clay wash.
2 Build up a substantial wax thickness. This can either be done by dipping successively in a large bath of wax, by pouring wax over the original, or by painting and building up the required thickness.
3 When the wax has cooled and hardened mark the sections of the mould, and cut through the wax with a sharp knife.
4 Make and place reinforcement to give strength and rigidity to the main mould and caps. The materials used to do this must be covered with wax to be held firmly in position. Allow the completed mould to harden.
5 To remove the caps soak the mould to break the suction between the mould and master surface. Carefully remove the wax pieces and take away the original. Clean the mould surface.

The mould is now ready to be filled with whatever substance is chosen; plaster of Paris, cement or resin. These materials are then cast.

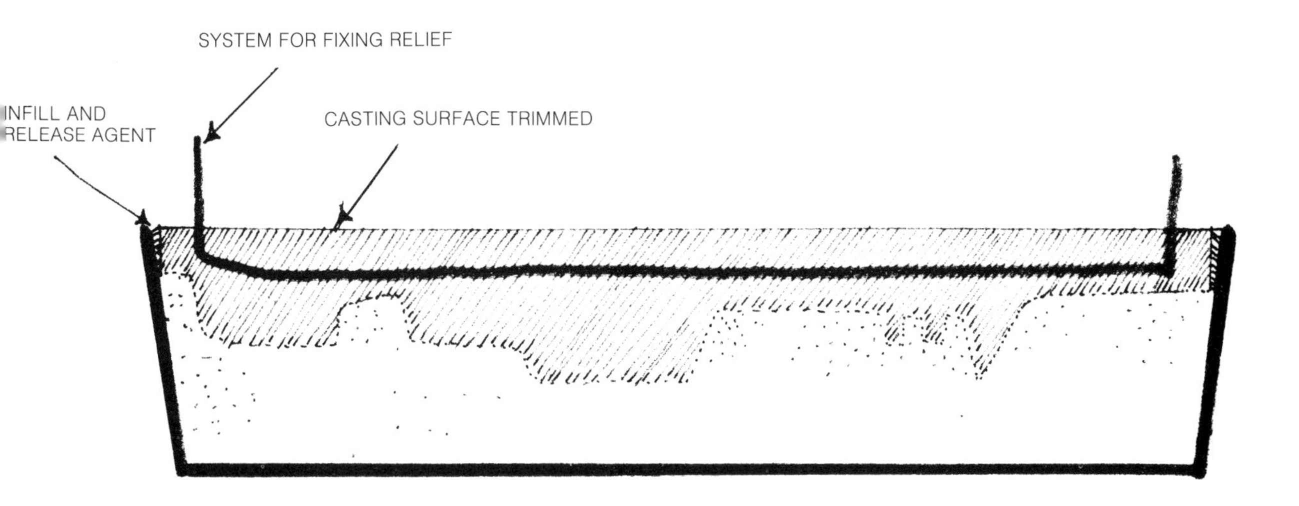

101 Sand negative showing the modelled sand negative and filling

Variations of this technique will obviously develop when experience is built up and a personal attitude has been gained towards the materials. Moulds can be built by casting slabs of wax and assembling them to make a mould negative, in the same way as clay or expanded polystyrene can be made to form a mould.

The technique for using wax is straightforward. Melt it, then apply it, build it up, reinforce it if necessary; remove it, clean it and fill it. A too brittle wax may inhibit the making of a mould. The wax needs to be pliable, yet firm, and able to hold its form. The parting agent between the wax and master needs to be simply a wet surface or a clay wash. Oil or a greasy surface will cause the wax to stick fast, as will a dry porous surface. Remove the pieces of mould carefully so as not to distort them or the main mould. Care should be taken when putting the mould together again and sealing the seams. Wax by its very nature will make a good release agent from most casting media. To be certain of proper release, however, it is not too cautious to apply a film of oil, liquid wax or polyvinyl alcohol, according to the filling.

Sand Ordinary builders' sand provides the sculptor with a material that can be used to good effect. By pressing items into damp sand, not too wet, an impression is retained to make a negative or female form, from which a positive or male form can be made. The sand is placed in a suitable container, a box or tray. The proposed design can be made in this by pressing, modelling and building up, (*101*), to achieve the negative mould of the designed form. The casting media can be poured directly into this mould, allowed to set and harden, after which the cast can be removed from the mould and the sand cleaned off. The kind of surface achieved in this way is, of course, granular, and work done on the cast surface can enhance this quality. Many sculptors use this method for making a quick maquette for relief sculptures. A variety of ideas can be tried quickly and looked at sympathetically in relation to the final form by employing a sand mould.

It is possible to make very large scale reliefs in this way. The frame retaining the sand can form the link between a number of panels, being included itself in the casting, together with any necessary reinforcement. All kinds of fixing can be placed during the pouring process, becoming an integral part of the structure.

The materials that can be cast from such sand moulds are limited to those that can be easily poured, i.e. plaster of Paris; concrete in a very large mould well packed; resins, if of a low viscosity – sticky high viscosity resins may drag down the surface; and metals, although it may be necessary to use a foundry sand to get the best results. Lead can be dealt with easily in the studio, but when pouring it, make sure the sand is just damp, not wet. The lead can be melted in a suitable crucible or cast iron pot.

The process of making sand moulds is, of course, open to innovation, adaptation, and

102 Aluminium relief designed by Geoffrey Clarke for Castrol House, London.

improvement.

A large aluminium wall (*102*), designed and made by Geoffrey Clarke for Castrol House, London, is one example of the extreme potential of this technique. An effective and interesting casting has been achieved, direct and simple in execution, by exploiting a particular character and form.

Lead Primitive or unsophisticated lead sculptures can be made by pouring the molten metal into a clay mould.

The clay can be modelled directly to make the negative form or be taken from a hard original (see page 28). The mould should be of good quality clay, having some added refractory body. It needs to be fairly thick in construction, having to withstand the weight and thermal shock of the molten metal. The clay can be made up and used again after use as a mould in this way. When the mould has been made it must be allowed to dry to a near leather hard state. This can be speeded up by placing the mould near a source of heat. When it is fairly dry and hard the molten lead can be poured in. This should be done in a steady flow. Do not stand over the mould when pouring the lead; excess moisture will cause the lead to spit and splash. Pour by means of two cast iron ladles, one pouring the lead into the mould, the other pouring lead from the crucible or melting pot into the first (*103*). In this way a continuous flow is achieved at a fairly safe distance. When the lead has cooled the mould can be broken away. The cast is ready to be chased, cleaned up and finished.

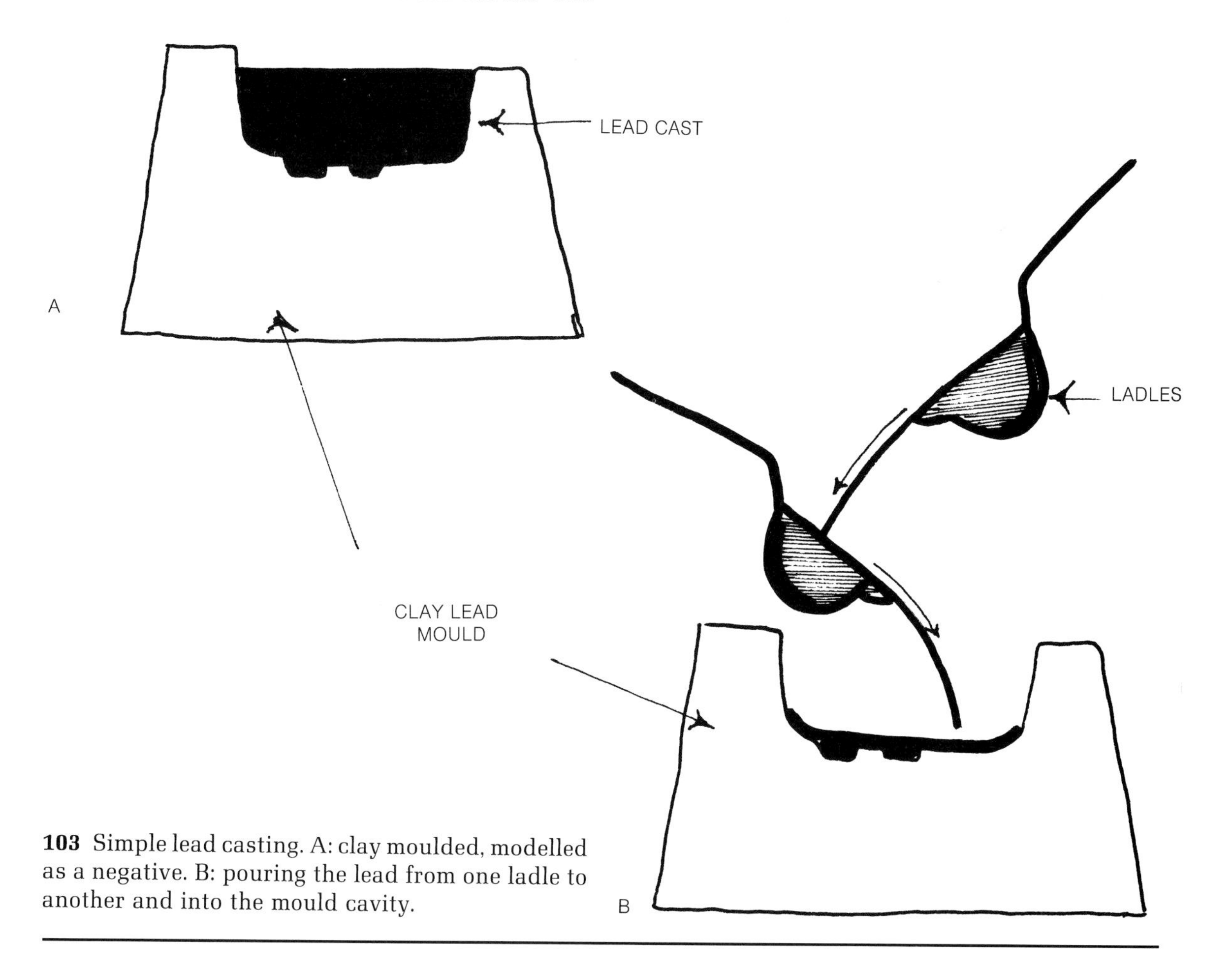

103 Simple lead casting. A: clay moulded, modelled as a negative. B: pouring the lead from one ladle to another and into the mould cavity.

V Metals

The business of making sculpture of metal has preoccupied sculptors since the discovery of melting metals in the Bronze Age. The mixing of metal, from early refinements of smelting to techniques typical of the twentieth century, occupies the minds of sculptors today. No matter how many new materials are available to sculptors, materials with fine integral qualities of colour, strength and durability, the metals prove to be the most compelling and rewarding.

The metal most connected with sculpture is bronze. Bronze casting becoming thereby an important ancillary to the art of sculpture. Today, of course, aluminium must be added to the repertoire of cast metal used by sculptors, as indeed should cast iron. Iron is not often used by sculptors, however, because of its extreme vulnerability to atmospheric corrosion. But since the Industrial Revolution, and the great innovation of cast iron as an architectural and decorative medium, it has been included in the inventory of metals available to sculptors.

The attraction of metal may be due to the resistance experienced from its surface, its cold ring, or may be the strange purity of colour, the density, or the empirical nature of the flowing potent molten metal and its dynamic, which forces an inevitable high regard. Perhaps it is the very old age of metal in the service of man which makes sculptors venerate it. Whatever it is that draws all sculptors at some point to work in metal, the positive force of the substance over synthetic materials will not be overlooked.

Young sculptors, often forced to work in less costly materials than cast metals, quite rightly do not contribute to the idea of the 'sanctity of bronze'. In fact some imply that they are 'put off' the material because of its history and tradition. The image, if successful, justifies the means; a fact which cannot be ignored. But I am inclined to feel this is because of the high, even exorbitant, costs of metal casting. This may be a good thing in the long run, in the forcing of further idiomatic change arising from the exploitation of newer materials. I am of the opinion, however, that experience in producing work in cast metals is invaluable to the sculptor, if not a necessity.

A bronze casting can be produced from almost any original. The various methods for producing an original or master cast, by one casting technique or another have already been described. The majority of sculptors take the master cast to a foundry; the foundry craftsmen being entrusted with the task of reproducing that master faithfully in the chosen metal. An alternative is for the sculptor to set up his own studio foundry and make his own castings. Some sculptors prefer to give the work to a professional foundry, demanding complete accuracy, but this may prove to be an expensive business, because of the skill required and the time it takes to produce high quality work. The sculptor, however, is freed from the headache and the backache of foundry work and can concentrate his efforts in new sculptures, but he must have an understanding of

104 *Station of the Cross* by Giacomo Manzu, 1950, bronze. *Kunsthistorische Musea, Antwerp*

the techniques involved in order to be certain of getting what he wants.

In the last 20 years there has been a veritable explosion of bronze foundries in the United States, and they can now be found right across the country. To some degree a similar phenomenan has taken place in Canada and Britain, where many more art bronze foundries have opened and produce fine work. Costs as a consequence are variable: it is possible to get a number of quotes to cast a bronze and so choose the one that best suits your budget. The possibility of casting in other copper alloys, aluminium, stainless steel and even iron has also increased with this renewed foundry activity. Because of the high labour content however, both skilled and semi-skilled, casting costs are still high, if not in some cases exorbitant, a factor that is unlikely to change. These costs when reflected in final pricing can often inhibit sales or commissions, especially for the student or little known sculptor.

Italian foundries no longer provide the cheaper, high quality alternative, we all enjoyed until the seventies. German foundaries have not sustained their myth for fine art bronze casting popularised by Henry Moore, but home choices are much greater.

Casting time is always estimated in months rather than weeks due to the pressure and work load of foundries. It is wise always to establish a good working relationship with the foundries you do use to gain mutual respect and understanding, which is more likely to ensure satisfactory results and put you in a position to negotiate prices. One solution to exorbitant costs can be to produce a wax in the studio, and take it to a foundry willing to carry out the subsequent processes.

The most safisfactory solution to such problems as foundry prices and time lag is the setting up of a foundry by a group of sculptors, each person contributing to its installation and to the production. It may, in time, even be possible to employ and train an assistant. Such a group practice would help to spread the pressure of the work; the furnace and kiln being run only when sufficient moulds are prepared, and each sculptor would be responsible for his own moulds, the firing and pouring being done together. A foundry immediately available is a distinct advantage but it must be pointed out that there are also disadvantages. It can be a time-consuming business that has to be tackled with care and caution and it is often felt that the time taken would be better spent making new sculptures. The control of a sculpture, however, from conception to presentation, is very desirable and does in fact give great freedom. The bonus is being able to finalise a sculpture in metal, and this is an important fact. This applies particularly to the young sculptor as more notice will be taken of work if it is in a material that is bought, used and generally regarded as presenting a definitive statement.

Metal castings may be solid or hollow. The greatest accuracy regarding volume and density of materials is best achieved by casting hollow forms. The wall thickness should be not more than $\frac{3}{16}$in to $\frac{1}{4}$in (4mm to 6mm) according to the size of the cast form. Solid castings although quicker to make, suffer in volume and accuracy because of the shrinkage of the metal upon cooling. This shrinkage is in the region of $\frac{3}{16}$in (4mm) to every 12in (300mm) of surface area.

Hollow castings are made, therefore, by either the lost wax or the sand moulding process for the following reasons:

1 Economy: less metal is used to make a hollow casting.
2 Weight: this is obviously less in a hollow casting than a solid cast. This might prove to be a crucial factor, structurally, in particular sculptures.
3 Strength: the hollow form is structurally the strongest.
4 Shrinkage: all castings shrink on cooling.

105 and **106** An industrial foundry furnace being charged with fresh metal ingot. The foundryman is wearing heat-proof gloves to protect his hands.

107 Aluminium sand cast sections in the foundry. Flash from the mould seams can be clearly seen plus the system of runner. The bosses on the back surface of the panels are made to be tap drilled for fixing *in situ*.

108 Discussion with the foundryman to determine the sections and sizes to suit the moulding process; in this case, sand moulding for aluminium.

This may cause cracking between unequal forms and it will certainly cause volumetric inaccuracy. Overall shrinkage can be very great, varying from metal to metal. The shrinkage can be controlled by making an even wall thickness and a hollow cast. This reduces the overall mass of metal and the consequent shrinkage.

The processes already mentioned are those principally used, the lost wax or cire perdue, and the sand moulding processes. A sand mould is a casting produced from a mould made from sand capable of standing up to the thermal shock and weight of the molten metal. The lost wax method is possibly the more complex of the two. The form is made at some stage in wax, which is then covered with a refractory mould. The mould is heated to melt out the wax leaving a cavity to be filled with the molten metal; the moulds in both processes being broken away from the metal when it is cool enough.

Both these processes have been used by man to produce items of utility and veneration since the Bronze Age. The principle initiated then is very much the same today as it was thousands of years ago. There have been very few innovations in this sphere of use to the sculptor.

Machinery and equipment, allied skills and materials necessary to industry, have developed from the basic crafts of bronze casting. Those concerned mainly with mass production have, of course, taken machinery and allied techniques into quite another sphere of metal working. Resin bonded sands, heated patterns or hobs, and pressure casting from steel hobs are cases in point. Industrial lost wax processes are recent innovations, but unfortunately the casting, although of extremely high quality, depends for low cost on a great number of castings being made. Usually, too, the size of casting possible is relatively small. For precision casting of large complex forms, requiring only 'one off', or at the most six castings, the skilled craftsmen working the traditional processes is the most economic and satisfactory.

Presenting the master cast to the foundry is perhaps a good point of departure in explaining casting techniques, whether the foundry be the studio or the professional establishment. The original should be of a hard material, permitting it to be handled by the foundry man and retained as a pattern. A good foundry moulder should be able to make a mould from any master cast. It may, however, be necessary to cut the master, or construct roman joints, to allow the master to be split up into components for casting. Such divisions should be planned and discussed with the foundry man. A good moulder will try to cut the master as little as possible, saving work later. The necessity for cutting varies from foundry to foundry. One reason is the size of the crucible available, an important factor too in the studio foundry. It is pointless to make a mould in which the enclosed volume is greater than the capacity of the crucible. Another reason for cutting is the manageability of the invested mould. If this is too large and heavy to handle easily, it may be damaged. Damage is a particular hazard when the mould is removed from the kiln and placed ready to receive the molten metal.

If a master cast has to be cut at all, it is best done in the sculptor's studio during the manufacture of the master. At this time proper roman joints can be made and checked by the artist, each piece being made to fit snugly and securely in position. The foundry man will visit the studio, providing he is getting the work, and advise on the divisions in relation to the particular foundry facilities.

Wax originals prepared by the sculptor either by modelling direct, or by making a mould to produce wax castings, can be taken to a foundry to be invested and cast. Some foundries do not mind doing this providing the artist accepts responsibility for any short run, or other mishap that may occur. The wax can be supplied with a core, pre-determined or poured into the hollow wax. Make sure that the foundry are willing to take on the sculptor's own wax. If it is cored, they must know what the core materials are and agree to use them.

Of course it is important to make hollow metal castings, whenever the factors such as shrinkage and weight are important. Very small castings can, however, be made solid quite easily. I often have large slender forms cast solid, when the amount of metal shrinkage is unimportant. The quantity of metal saved in making a small cast hollow is negligible. The effort involved in making a hollow cast in these cases could be better used in making forms, assessing shrinkage and juxtapositioning compatible volumes, distributing them more equally to avoid fractures. This can be done to some extent by taking care in placing and selecting the size of runners. If a form is too great in volume and cracking upon shrinkage predictable, a solid casting can be made by introducing an armature. This armature when introduced at the centre of the larger volume, acts as a chill bar. The chill bar causes the larger quantity of molten metal to cool at a similar rate as other parts. In this way the shrinkage is not so violent, reducing the possibility of fracture, due to unequal shrinkage.

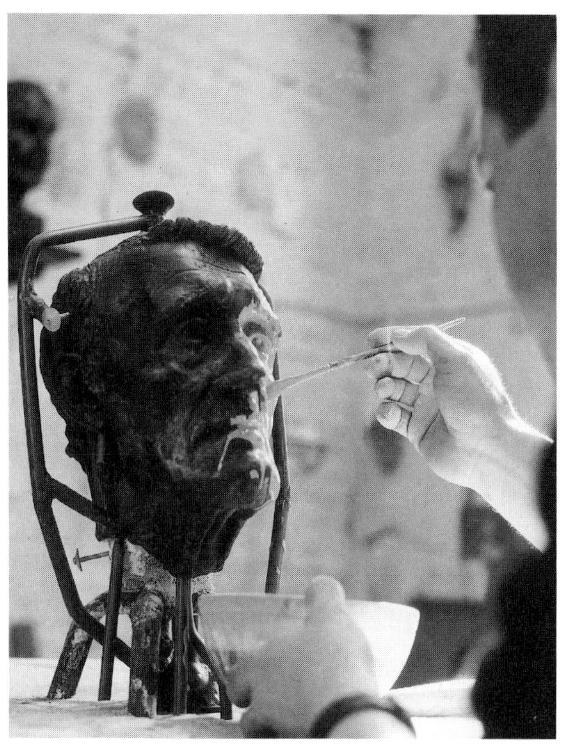

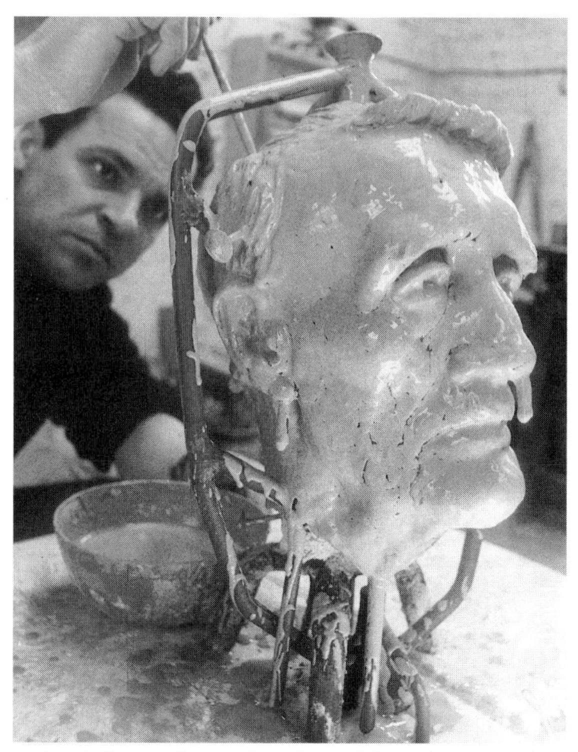

109 and **110** The wax positive has been filled to make the core and is provided with runners and risers. The thick rods are the runners, the thinner rods at the neck are the risers. Note the core pins. The first application of investment mould is being applied with a brush.

Lost Wax

The following description is presented as an itemized procedure. By this means the various stages are explained together with the various terms. The effort involved in producing a metal cast makes it necessary too, in my view, to cast a form worthy in the sculptor's mind of that effort. The basic forms in this book therefore although they would provide useful exercises, may seem unworthy of the effort. I will, therefore, give diagrammatic illustrations using them, but leave the choice of the forms for exercise up to the individual. The lost wax casting processes are carried out as follows:

1 *The wax original* is prepared. This done by moulding from a master cast, usually by means of a flexible moulding compound. From such moulds wax castings are prepared. The sculptor may produce his own wax for the foundry. If the wax is prepared at the foundry, the sculptor will be asked to check it against the master. Any touching up, cleaning off seam flashes, signing, or inscribing the edition number, is done at this point. Bronze sculptures are often produced in small editions, usually six in number, sometimes more. Each casting must have its edition number marked on it, the number of the particular casting shown over the total edition number, e.g. $\frac{2}{6}$. The edition allows casting costs to be spread over the whole series.

2 *Core* material is poured to fill a hollow wax. This usually made up of 1 part plaster to 3 parts grog (ground ceramic). Before filling the core, weigh the wax to determine the amount of metal to be used. The ratio of 1lb

111 As the mix stiffens the investment is built up by throwing. This consolidates the mould and speeds up the process.

(0.45kg) of wax to 10lb (4.5kg) of bronze is usual, plus the estimated weight of 'risers and runners'. Most experienced casters make this estimate by the size of the crucibles, which are manufactured to 1lb (0.45kg) capacities. The core material must be poured to fill the wax completely. If a piece mould or flexible mould is used, it is convenient to pour the core while the wax is held in the mould. It may be possible to replace the wax in the mould, after weighing it, to pour the core. This will prevent any damage to the wax, when vibrated to release trapped air bubbles, and to make a well consolidated core. A core vent should be placed at this point (*116A*). Large cores require a substantial core vent. Smaller forms can be vented by placing a waxed string, held firm by plaited wire, allowing it to be placed conveniently in the core (*116A*). This system was devised by a colleague, Michael Gillespie, who in his own foundry produces some of the finest small bronzes. The core should extend beyond the wax (see pre-determined core page 131).

3 *Iron pins* are tacked into the core, through the wax, with sufficient projection from the face of the wax to be gripped by the mould (investment) (*116A*). These pins placed at strategic intervals about the work ensure that the core and investment stay in the same relative position when the wax is melted out. If there were no pins, obviously the core would move about loosely once the wax had gone. This would result in the passage of the molten metal being blocked, and the casting being a failure.

4 *The pouring gate* has to be designed and fitted. This is the name given to the system of channels through which the molten metal flows and gases escape. The metal flows via the pouring 'funnel', through the 'runners' into the mould cavity. The runners extend from the funnel to carefully placed points on the sculpture to ensure the quickest and

112 and **113** A linoleum container is made and placed over the invested wax.

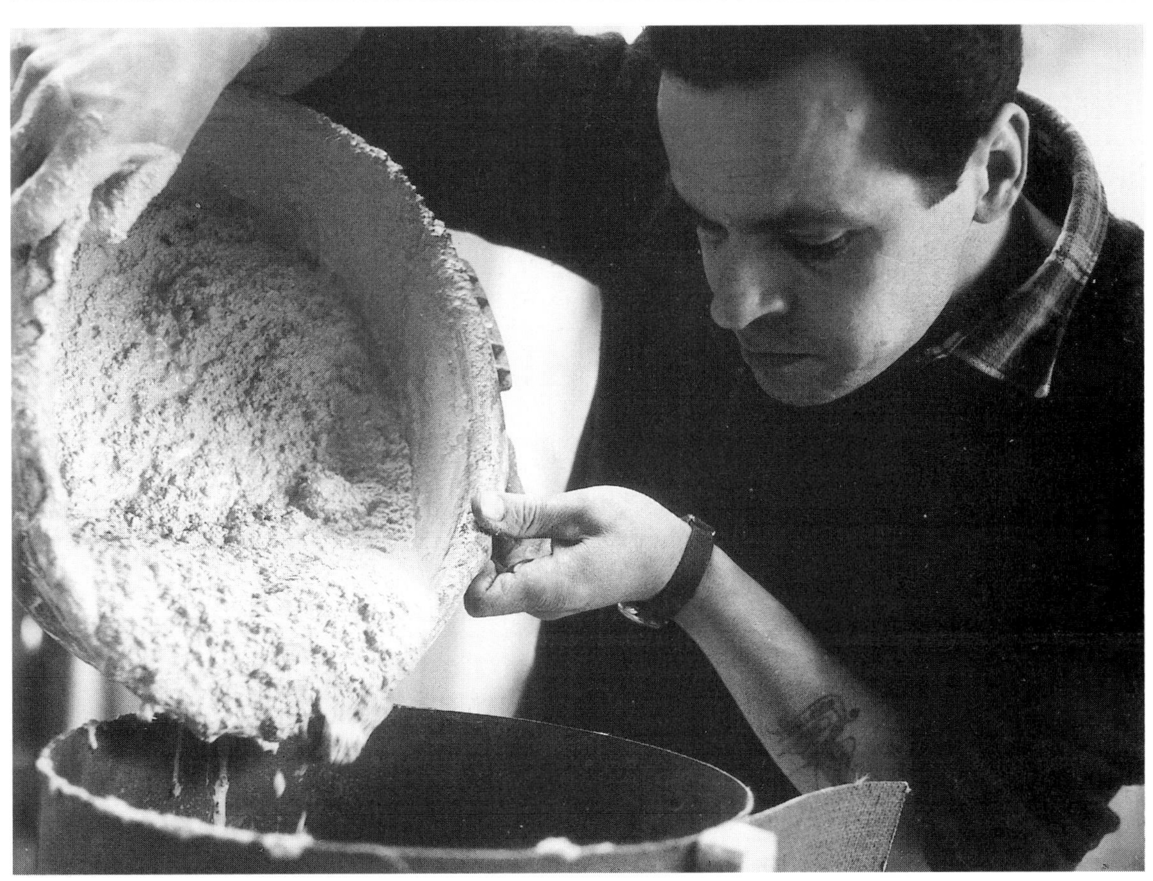

114 (*Top*) A coarse mix of grog and plaster is poured to build up a strong refractory mould.

115 (*Bottom*) The linoleum is removed and the pouring gate ends revealed. The large opening is the funnel, the next are the core vents and the smallest are the risers.

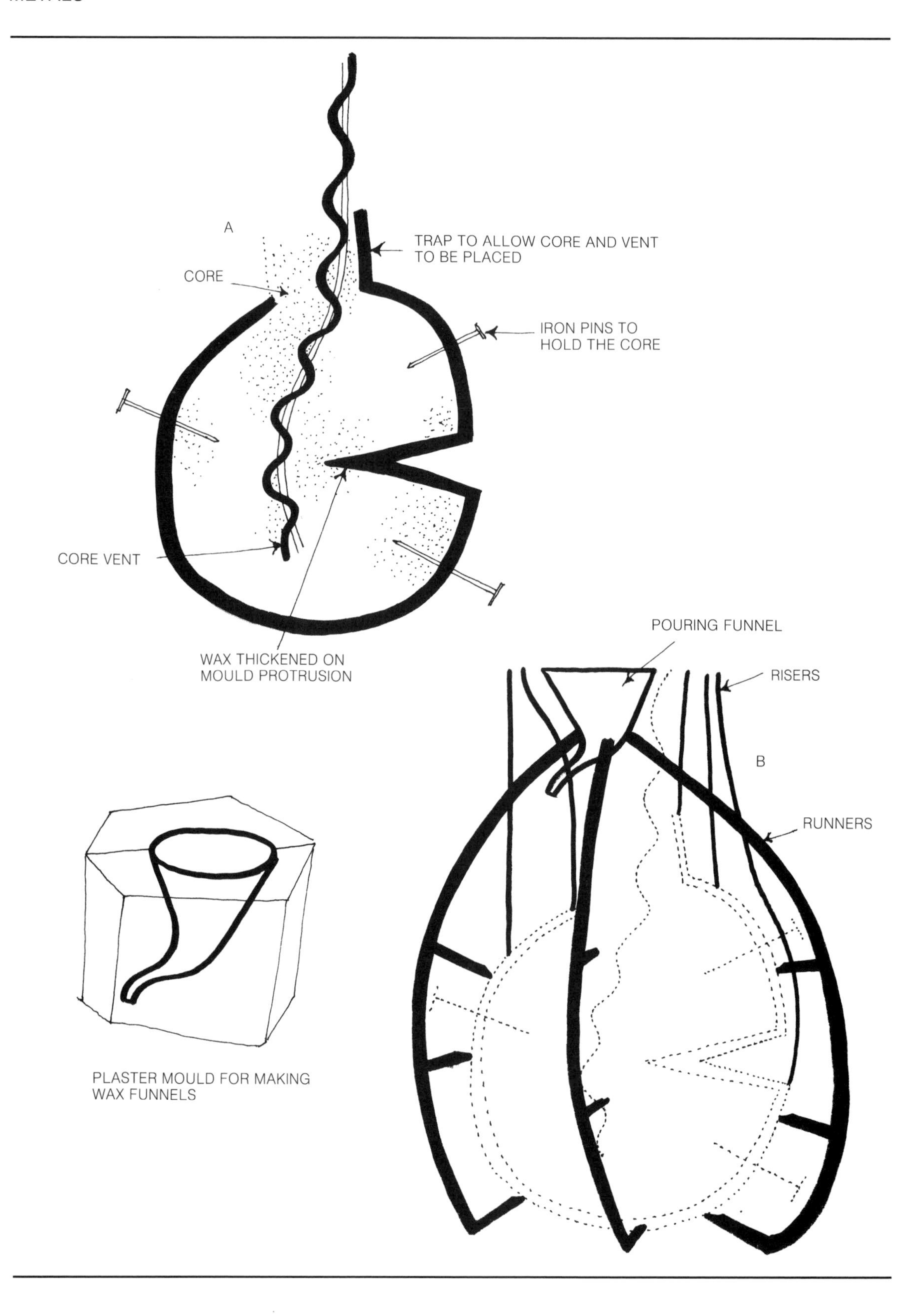
A
TRAP TO ALLOW CORE AND VENT TO BE PLACED
CORE
IRON PINS TO HOLD THE CORE
CORE VENT
WAX THICKENED ON MOULD PROTRUSION
POURING FUNNEL
RISERS
B
RUNNERS
PLASTER MOULD FOR MAKING WAX FUNNELS

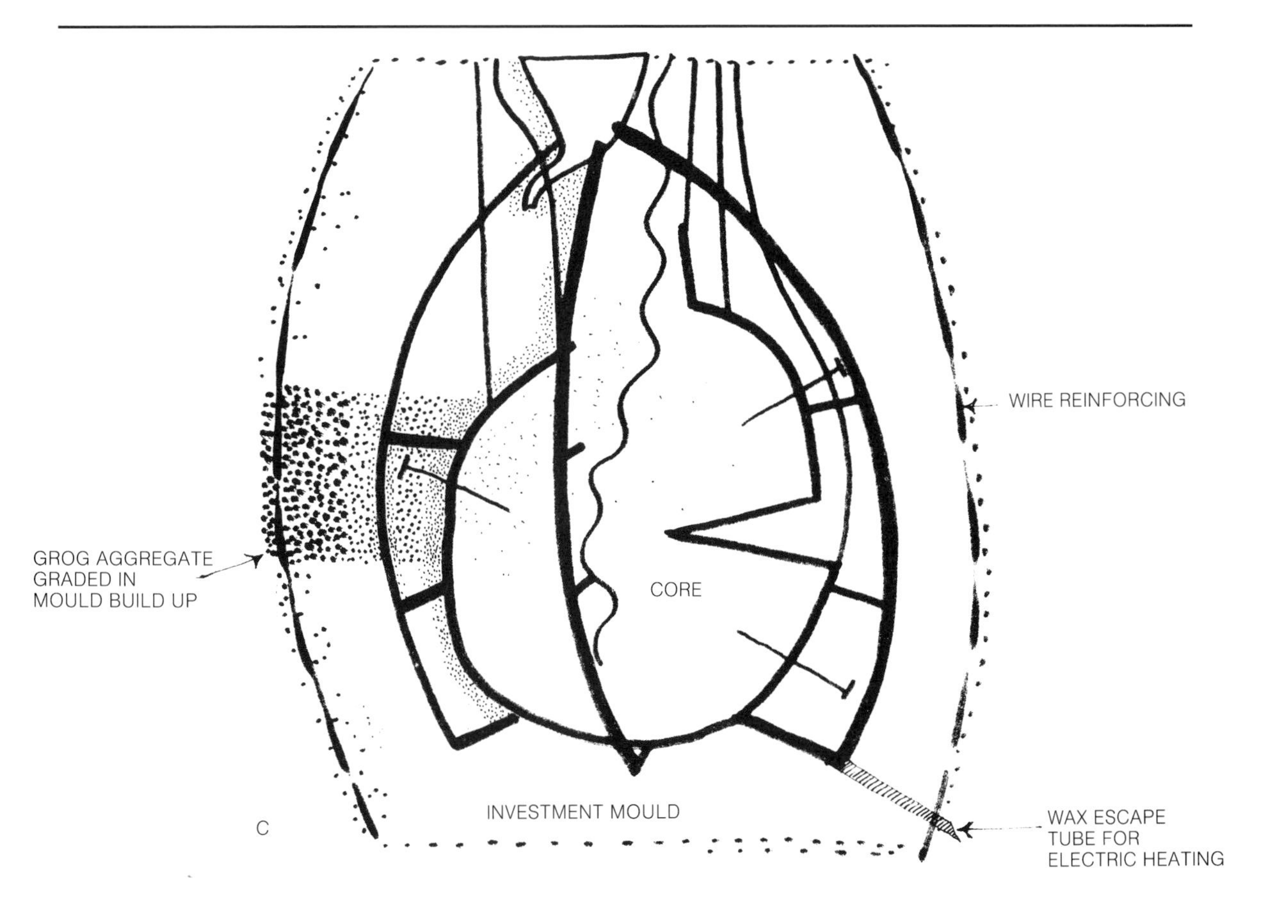

116 Various processes involved in the lost wax technique. A: the wax thickness, core and vent, and iron pins to secure the core and investment. B: the system of runners and risers, including the pouring funnel. C: the investment mould and the grades of grog varied through the mould build.

smoothest flow of metal. Several runners are usually necessary according to the surface area of the cast. The metal must run quickly to every part of the mould cavity without chilling. Gases created by the metal run before it, driving out air from the mould at the same time. These gases and air escape up the 'risers' which are vents taken from those features of the positive form likely to trap air, and prevent the flow of metal to the finer forms. Air traps will normally occur at undercut forms, protruding from the main form, when this is in position to be filled.

Runners and risers are made of wax rods, varying in thickness, and are fixed to the wax casting by simply applying a hot metal tool between the two, welding them together. In this way the pouring gate is built up (*116B*). Runners are thick rods, the diameter varying according to the size of sculpture, from ¼in to ¾in (6mm to 19mm). The larger the casting the larger the runners. Wax rods can be easily made, using a wet two-piece plaster of Paris mould, from a wood dowelling pattern. Some sculptors cut strips from wax sheets to make runners and risers. The risers are usually of a narrower gauge being placed simply to allow gases and air to escape. On small forms risers can be made quickly by dipping string into the molten wax. The funnel which surmounts the pouring gate can be made by modelling a funnel, and making a simple two-piece mould of plaster (*116B*). This can be used thereafter to make wax funnels as required. A paper funnel can be formed by simply folding the paper then dipping it in wax. Another common useful method for making a pouring funnel or cup, is simply to use a manufactured paper or waxed paper drinking cup. Once given a coating of wax by dipping or painting wax on, these can easily be attached to the runners, but make sure such attachments are well made (*116*). A trap should be placed over the funnel opening to prevent dirt or foreign bodies entering to foul the mould cavity. The funnel can be stuffed with newspaper which will eventually burn out, or be covered with a waxed paper plate, as in figure *116*.

5 *The investment* or mould is now placed around the finely prepared wax, which has been cored, pinned and fitted with a pouring gate. The investment can be a specially prepared refractory material developed especially for the foundry industry. Advice regarding such investment materials should be sought from the suppliers. An investment can be made with plaster of Paris mixed with grog (ground ceramic). The grog should be sieved to make three grades; *fine*, to mix with the first application, *fairly coarse*, for beginning the mould build up, and *very coarse*, to complete the mould. This mixture of materials is still used by most foundries in preference to proprietary investments. It is obviously cheaper in large quantities.

To achieve reproduction of the finest detail the investment is carefully applied to the wax surface. This can be treated first with methylated spirit to remove any grease. Paint the initial layer of mould with a very soft brush. Do this quickly over the entire surface, including the pouring gate. Be sure to leave a well-keyed surface to make the best adhesion to the next layer. The mould after this initial surface must be built up to make a substantial thickness, which by its nature maintains a maximum strength when baked (*116C*), the sieve size of grog increases as the mould grows. The final application should be reinforced with wire to gain extra strength. Some alternative methods for placing the investment are by pouring or dipping.

Both methods can speed up an investment and are particularly useful for small to medium sized pieces; that is, sculptures that can be comfortably handled. Containers are well constructed to be filled with investment made more liquid than that used to build up an investment mould. Into these containers the prepared wax can be either fixed cup down and the investment carefully poured around it, or the container can be filled and the prepared wax dipped into it to be held in place as the investment hardens (*116*). The containers are called flasks and can be made up of linoleum, tarred paper, rolled metal or tin cans. Metal flasks are manufactured for jewellery casting sometimes with a cast rubber cap to seal one end of the tube to facilitate pouring the investment. These flasks are in fact precisely cut open ended tubes of various sizes made to fit either a vacuum or centrifugal casting system, but the principle can be usefully employed on larger casting projects for sculpture. Metal flasks can be left in place during the burn out and the pour and will give added strength to the mould and hence greater security. If tin cans are used as flasks they too can be left to give added strength to the baked mould, but the bottom of the can must be pierced before the mould is placed in the kiln to bake to allow moisture to escape. If this is overlooked, steam will build up and cause an explosion within the can. All other flask materials should be removed preferably just after the investment has set and before it hardens and expands to avoid cracking in the investment. According to the size of the mould, reinforcing should be made integral to the mould on page 125. Wire reinforcing is particularly necessary on slender moulds that can be easily broken as they are being removed from the kiln and turned to be packed ready for receiving the molten metal.

6 *Baking* the completed investment is the next process. This melts out the wax to form the mould cavity and also dries out the mould thoroughly, thus preparing it to receive the molten metal. The baking process is best begun when the investment is fresh as this helps the wax to drain quickly. The mould is baked in a kiln fired by gas, oil, wood, coke, coal or electricity. Any fuel may be used as long as the heat can be built up fairly slowly and maintained at a red heat. The traditional method is to build a kiln around a number of moulds over a fire box. This kiln is dry built of refractory bricks (*117*). Another method is to heat the mould by binding about it an electrical resistance wire connected to a variable resistance control. In this way a controlled heat can be built up to bake the mould. Over an open baking system, the mould can be simply turned upside down to bake, and the wax will run out from the funnel. Electrically heated and baked moulds require packing in an insulatory material such as *kieselguhr*, and cannot be easily moved. Therefore it is baked in the pouring position, the right way up, allowance being made for the wax to run out. A satisfactory method is to make one of the runners extend through the investment at the lowest point. The wax will drain from this, through an opening in the insulator which should be made with a metal tube. This opening must, of course, be plugged before the molten metal is poured, or else the caster will find himself standing in a pool of molten metal, which I do not recommend. This drainage system can be employed to prevent the wax running onto the heating source in any of the other systems. Some casters like to collect the wax as it drains out; in this way some check is kept on the amount remaining to be fired out.

The *Kiln* can be electric or fired by some other fuel. An electric kiln can be the same kind as those used in the ceramics field. This should have a large firing chamber, however, capable of containing large cumbersome moulds. I would suggest that a chamber 4ft × 4ft × 4ft (1.2m × 1.2m × 1.2m) minimum should be specified, if the kiln is being built to order; smaller firing chambers will prove quickly inadequate. Of course, the largest possible kiln is preferable, but cost is a deciding factor. Electricity provides the most easily controlled baking facilities with a minimum of effort, once the kiln is installed. Traditional systems for baking are to build the kiln over a fire box to fit the mould or number of moulds. The fire can be a fixture over which a kiln of almost any proportion can be built. The kiln is built up of high fired refractory bricks. These can be held together with a mortar mix of clay slip and sand, placed at intervals only, not between every brick. Unnecessary holes in the structure can be plugged with this mixture. Do not fill all the spaces between the bricks because they help to aerate the fire and maintain a good red heat. Leave a space at the top to charge the fuel (*117B*).

When stoked, the fuel should not be higher over the fire box than the bottom of the mould, or lowest mould, and it should not be allowed to clinker. This is particularly important when using coke which has a tendency to clinker quickly. The fire should be maintained at a steady red heat, sufficient to cook the contents of the kiln. This is usually achieved at between 24 and 36 hours. The invested moulds may not reach red heat; in this case they should remain in the kiln for a few hours longer. An old Italian saying amongst foundrymen is 'when the mould is cooked, cook him again'; just to be on the safe side.

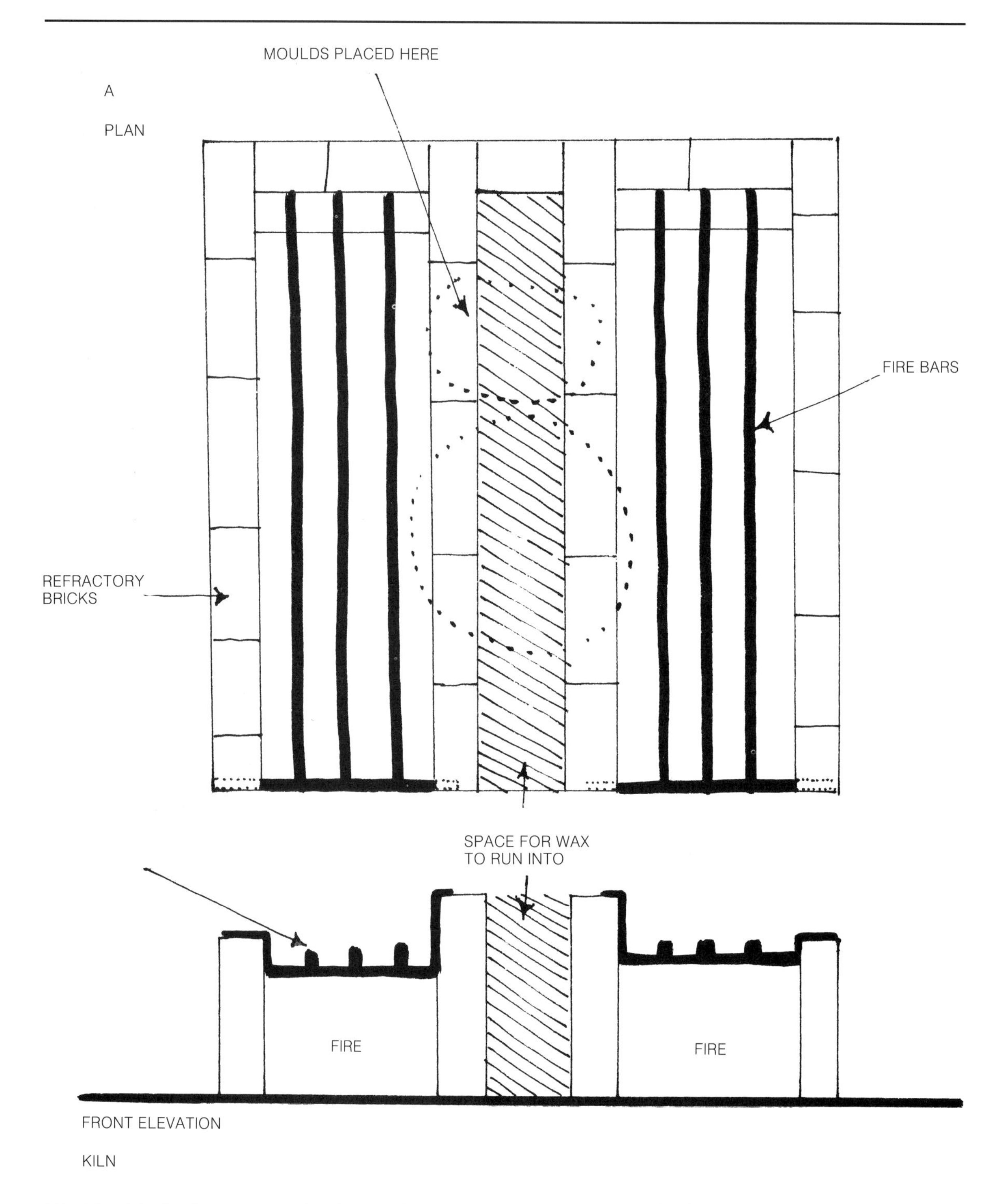

117 The kiln built up over a fire box to fit around the moulds ready for baking.

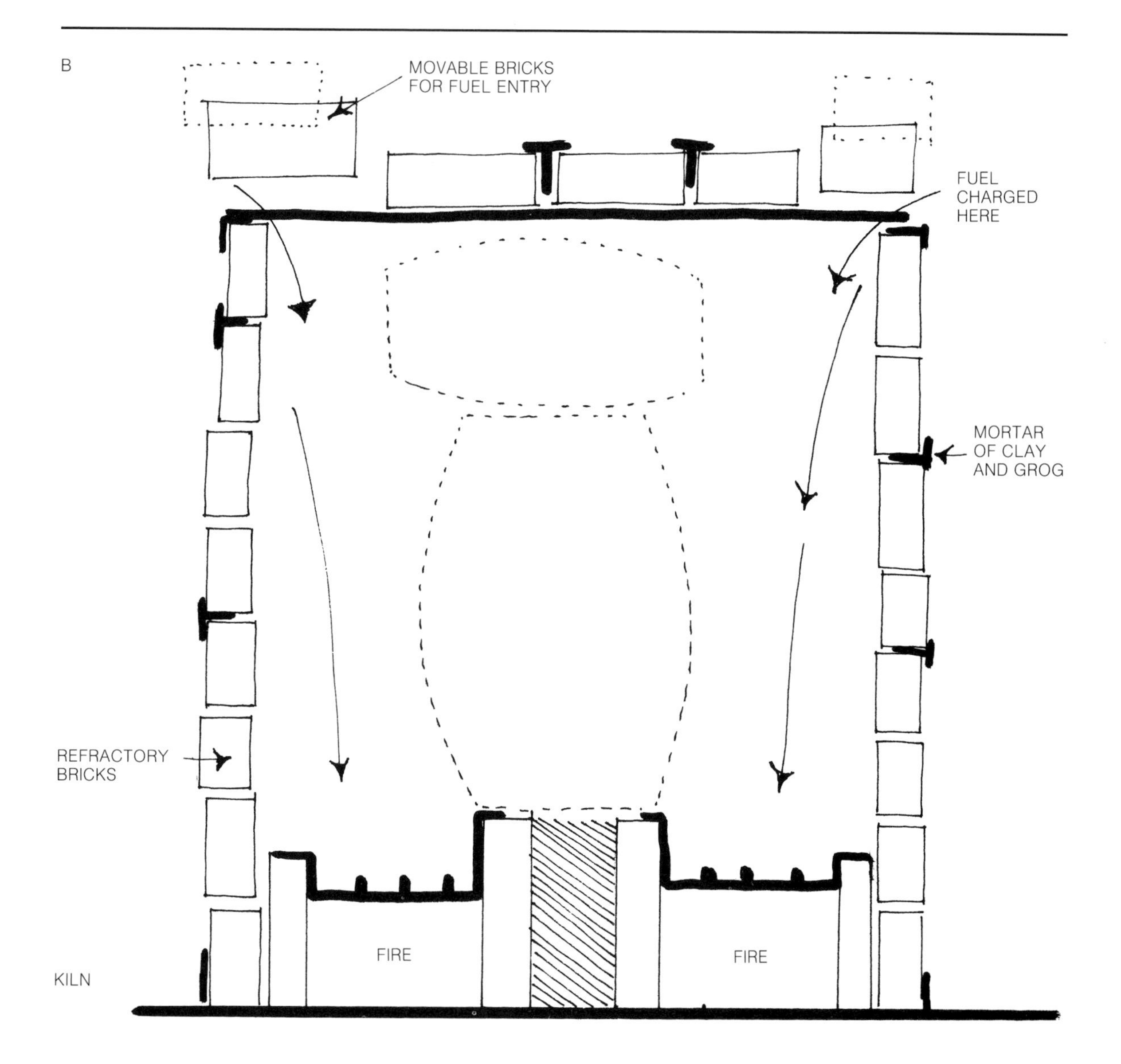

If the mould includes a resistance wire to bake out the wax, the heat can be calculated by the resistance of the length of wire and controlled by the flow of current from the variable resistance. The mould should be heated gradually to red heat.

The invention of ceramic fibre (alumino silicate) has made the making of kilns and furnaces much simpler and an even more viable proposition for sculptors to make their own according to their particular requirements.

Many variations on the theme are possible, of course, but all depend on a proper understanding of the traditional principles and requirements for burning out the mould and melting the metal accurately and safely. Do not take risks when baking moulds or melting any kind of metal.

7 *Check* that the mould is properly baked. No wax or moisture should remain inside. To achieve this degree of readiness, the moulds

must reach red heat, or as near red heat as possible, all over. They must be maintained at this heat for long enough to burn out the wax and any carbon deposit, and to drive off any moisture from the investment. Although the main body of wax will melt and quickly run off when heated, small amounts may be absorbed into the porous investment, and must be burned out. This is the case when it is not possible to bake a freshly made mould but have to wait for a number to be prepared. Visually, the properly baked mould will be free of smoke and steam, and if the funnel can be seen it will be free from any carbon deposit. The mould should remain at red heat until this happens. A cold mirror or piece of polished metal held over the mould will help to detect any moisture which must be eliminated.

8 *Packing* the baked mould, to hold it firmly in place, is the final process before melting the metal to pour. In the case of moulds baked by means of the resistance wire, the insulation packing is sufficient. The drain hole must be plugged, then the metal poured (*116C*). Other moulds baked in a kiln need to be packed into damp sand. They are placed in an open box or pit, then the damp sand is packed firmly around them till only the funnel and riser opening are visible. These openings should be covered to protect them and to prevent sand dropping into the mould cavity. The packing ensures the position of the mould for pouring and prevents the mould cracking and the metal running out. Before packing the mould, some casters place a reinforcement of scrim and plaster to prevent cracking. The metal should be prepared during the packing operation (see furnaces and melting metals, page 143).

Ceramic Shell

The procedure I have described for casting bronze applies to that bronze casting process now referred to as 'traditional investment casting'. Industry has caused another investment material to be available, known as ceramic shell. This is in fact a moulding process that has nothing to do with ceramics, but is a material that allows an investment mould wall thickness to be as little as $\frac{1}{4}$in (6mm), like a ceramic. This mould has great strength plus the ability to withstand thermal shock of high intensity making it able to receive the full impact of molten bronze easily without breaking.

The prepared wax, complete with pouring gate and cup is first coated with proprietary slurry made up from colloidial silica and silica flour. This coating is applied over the whole wax, which is first painted with a wetting agent to ensure good spread and adhesion of ceramic shell (slurry). The first layer picks up all the detail to provide an accurately cast surface. It can be made by painting or by dipping the wax, according to the size of the piece; once it has been evenly applied all over it is left to dry and harden. The work can be either left free standing or hung up in a even dry atmosphere; some makes of ceramic shell investment include an agent that changes colour as it hardens as a guide to progress. Subsequent layers of ceramic shell are then applied including a large particled fused silica, that makes a stucco built to achieve the required investment mould wall thickness. The fused silica can be thrown over the painted ceramic shell or applied by dipping the piece into a fluidising bed of fused silica. Each application must be allowed to dry and harden before another is applied. The covering must be even, erring on the thick side, but once the desired wall thickness has been made the whole thing must be left to harden. The wax then has to be removed.

Removing the wax is the next most important step in this process. This must be achieved quickly and evenly without the wax expanding to crack the mould. A microwave oven can be used, according to the size of the mould and the oven, but more often the de-waxing is done in a kiln or furnace pre-heated to above 600°C (1112°F), so that the wax is melted from its outer surfaces inwards. This prevents the expansion of the wax and cracking. Immersion in a bath of hot molten wax is another method of de-waxing.

Once the mould is completely free of wax it can be heated immediately to 1000°C (1832°F) and held at that heat to burn off any carbon. It is safe now to remove the mould from the heat and pour in the molten metal straight away. This can be done with the mould free standing.

Should the mould be too large to handle and support easily it should be packed in dry sand and never into damp packing material of any kind. This unique investment allows ventilation through its wall thickness, which means you can cut down on the number of runners and risers you use for the pouring gate, because the metal will flow with greater ease, and any air or gases will be partly evacuated through the investment. The porosity of the mould, however, means that it will ingest moisture, hence the reason for not using damp sand to pack around it for support, as is the norm with all other investment casting.

Although ceramic shell and its related technique is used to cast life size and over in art bronze foundries, for the studio founder it is most useful for small pieces using a pouring capacity of 50lb (25kg) of molten bronze. It is possible to make a quasi production line by dipping and coating small waxes as they are made ready, leaving them to harden off between layers. A hook device included in the runner cup will enable the work to be hung up to dry and harden ensuring a good circulation of air to aid the curing. Batches can then be de-waxed and cast, thus making good use of the process.

Should work need a core, this can be made from a traditional investment of grog and plaster, but this will increase the burn out time to get rid of the wax and to prepare the mould. This simply means doing these things a little slower than with pure ceramic shell so as to accommodate the older material. Hollow forms, such as a head, can be moulded with ceramic shell by cutting a section out of the wax to allow the slurry to circulate on the inside and the outer surfaces of the wax and so permit a mould build up on the inner and outer surfaces. The removed wax section can be cast separately and, after fettling and cleaning up, it can be welded back into place and chased to blend in with any surface treatment given to the finished work.

Pre-determined Core

Before moving on to sand mould techniques (see page 132), an explanation of the technique for making a bronze cast with a pre-determined core must be made.

This is in fact the method used by the more primitive societies, or sometimes by sculptors who prefer to model directly in wax, to make a unique sculpture (not one of an edition). The core is modelled first in a suitable medium, grog and plaster for instance. This form is taken to within $\frac{3}{16}$in (4mm) of the final surface, according to size, the final surface being built up in wax, on which any detail is made. The core, of course, is prepared with suitable core vents in readiness for the pouring. The wax original prepared in this way is subsequently prepared with a pouring gate of runners, risers and funnel, and with pins placed. The remaining processes, baking and pouring, are the same also.

This system of pre-determined core lost wax casting was used by the sculptors of the Renaissance. Benvenuto Cellini in his famous autobiography describes the process in some detail and with much drama. The bronze casting craftsmen of Nigeria, the Benins, produced sculptures of remarkable finesse and beauty using this technique (*118–112*).

118–122 Examples of the various stages in the production of a Benin bronze casting. *British Museum*

118 (*Right*) The wax original, moulded direct.

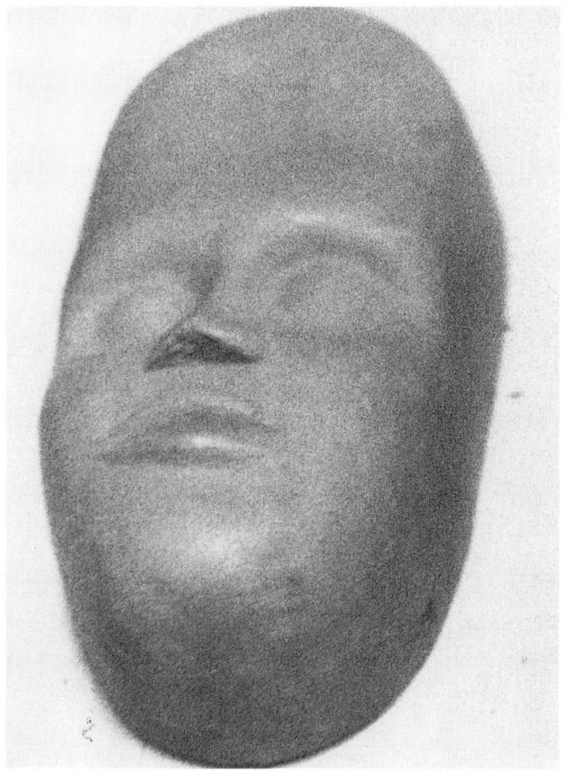

119 The refractory core or inner mould.

120 The completed wax and core.

121 The raw bronze casting straight from the mould. The funnel and runners can be seen.

122 The final bronze chased and polished. Decorative markings are included at this stage.

A refinement worth mentioning here is a lead armature over which a wax form is made. To make slender or delicate forms without a core, the lead can be beaten out and shaped to make a suitable armature. When baked during the processes of casting, the wax and the lead will melt and run out.

Other idiosyncratic forms in sculpture have been made in a similar way. If an investment mould is put around an organic or man-made object that will burn easily, a reproduction of that form can be made in bronze. A pouring gate can be affixed to the object prior to the investment. I have seen chairs, fruit, vegetables and insects reproduced in metal in this way. A friend of mine tested the quality of a new investment by casting in bronze the form of a beetle and achieved a remarkable reproduction.

Sand Moulding

Sand moulding is a means of producing metal castings from moulds made of sand. The resulting casts are coarse and granular in comparison to those produced by the lost wax process. This fact should be borne in mind when designing or contemplating using the technique. In the recent past, new kinds of sand, and binding agents used with the sand have been developed so that finer fidelity castings may be produced.

The original or master cast must be of a hard material, as it will have to withstand a certain amount of vibration and strain. Plaster of Paris is adequate; cement or polyester resins being harder make it possible to take a number of castings with little damage being sustained by the master cast. In the industrial foundry a

123 A large sand box, or flask, in the foundry. The second section has been put to the first and sand is being rammed up. Note the bars across the box and the metal pieces, or gaggers, that help to hold the sand, a necessity in a box of this size.

number of proprietary materials are used, manufactured expressly for the purpose of making durable foundry patterns (positive forms). These in general are plaster based but when cast and set have an extremely hard surface. Plaster manufacturers supply specially hard retarded plasters for this use. More specific information can be obtained from a foundry suppliers who know what is available in a particular area.

The original must be divided into suitable components to permit handling when being moulded. It is easier to mould small simple forms. As with the lost wax process, any dividing necessary is best done in the sculptor's studio, under his control; consultation with the foundryman of course being sought. For a description of foundry patterns made using expanded polystyrene moulds and formers see page 102.

The sand moulding process is as follows:

1 The master cast or section is placed on a bed of fine dry casting sand. This is contained in a *sand box* (*flask*) (*124C*). This box is a rectangular metal frame, which is made in various sizes. When over 14in (35cm) square the box includes bars across it to carry the sand. The master cast is placed in this and pressed firmly into the bed of sand, which is *tamped* down to form a firm sand surface around the master. The master and sand bed

are then dusted with a parting agent such as French chalk or silica dust.

2 The principle involved is that of piece moulding (see Chapter III). A piece of mould is made, making the necessary allowances for any undercuts; undercut forms requiring separate pieces of mould. Each piece needs to be trimmed carefully to fit properly one against or into the other, being tapered to draw, and dusted with a separator. The main body of the mould section can be made to pick up the greater part of the form. When drawn off the master cast the main mould must also draw off from the separate sections cleanly. This main mould can be, and usually is, designed to be the case, which contains and maintains in position the separate pieces. If, however, a case is necessary to hold a number of sections, it is made as follows: dust the sections of mould already made; place another flask on top of the first, and these can be clipped together. This is then filled with sand, which is tamped down to form a well consolidated case, or main mould and case. The sand is trimmed level with the flask edges, and tamping is then done with a rubber headed mallet. The consolidated sand is held firmly in place by its own pressure exerted outwards on the metal sides. The sands used in the foundry for moulding are generally excavated from quarries. These are naturally bonded sands, retaining the natural clay bond between particles, which permits them to maintain a modelled shape. Sand such as that found on sea shores will not bind sufficiently well to retain a certain form, because the natural bonding agent has been washed away by the salt water and, although of fine quality, cannot be used until it has been mixed with a synthetic binding agent. Such a synthetic bond can be achieved by adding fuller's earth, Benolite or oils; linseed, vegetable, fish or petroleum oils can all be used as binders. The clean sand and bond are blended in a *muller* or a large machine with revolving paddles and a mill. Oil-bonded sands are baked at fairly low heat to give them strength and rigidity. Seashore sand is usually finer in grade and can be packed to make a denser surface than *greensand* (natural bonded). Consequently after the addition of a binding agent, it can be used to produce finer forms and to make moulds requiring a larger number of pieces.

A further refinement is a sand called CO_2 sand, which has a synthetic binder and becomes hard when CO_2 gas is directed at it. This is the means used to harden the sand sections made on large forms that cannot easily be heated. It is also used to mould around an expanded polystyrene pattern. It can be placed with less tamping and therefore with less risk to the pattern. Tamping to consolidate the sand requires some experience. If it is overdone, the sand consolidates to trap gases and air which must escape. If too loosely compacted, the sand will not hold its form when the metal is poured. The latter case must be avoided, but with the former extra risers must be made to assist gases to escape and so produce a sound, dense metal.

3 When one side of the master has been properly completed the two flasks are inverted, and the sand bed removed to reveal the unmoulded side of the master. This must be dusted with the parting agent. The necessary pieces of mould and case are made in the same way as on the first side.

4 When the master, or a section of the master, is completely moulded and the mould firm, the positive form is removed. This is done by separating the flasks at the split line or seam, revealing the sand piece mould negatives which form the halves of the master positive. When the flasks are parted the original can be carefully drawn from the mould. The separate pieces of mould are then carefully removed from the master and pinned securely in position in the mould case. Long

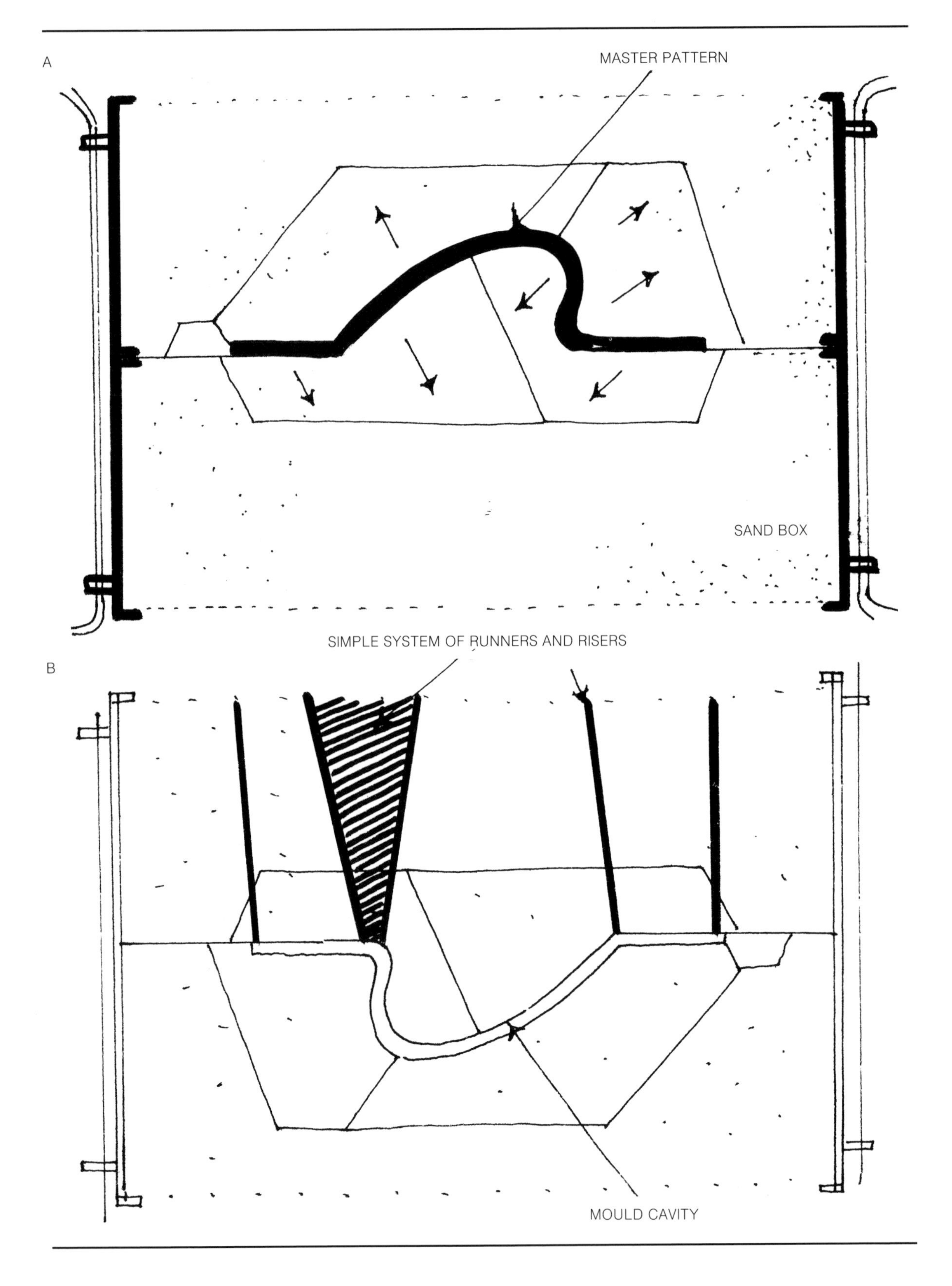
A
MASTER PATTERN
SAND BOX
SIMPLE SYSTEM OF RUNNERS AND RISERS
B
MOULD CAVITY

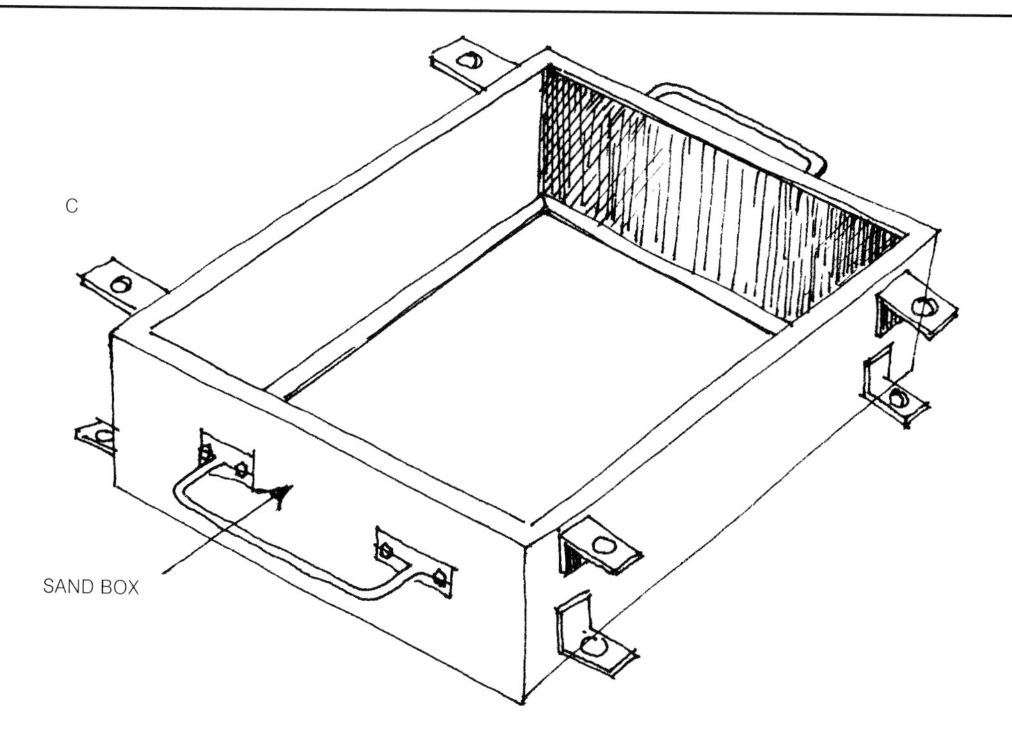

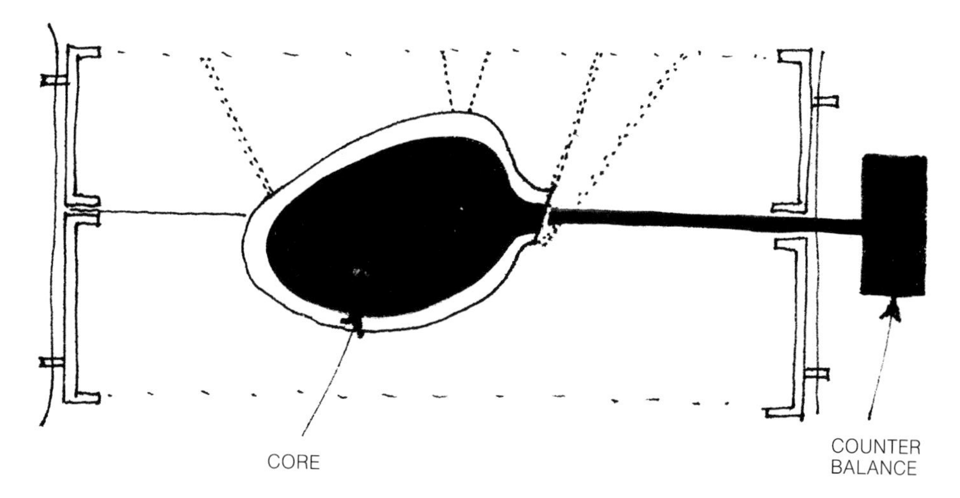

124 Sand moulding. A: the piece moulding principle employed to make a sand mould from a pattern positive. B: the mould cavity and a simple pouring gate drilled through the sand. C: a sand box or flask. D: a method of suspending a core in the mould cavity. See page 138 for diagrams E and F – a more complex pouring gate.

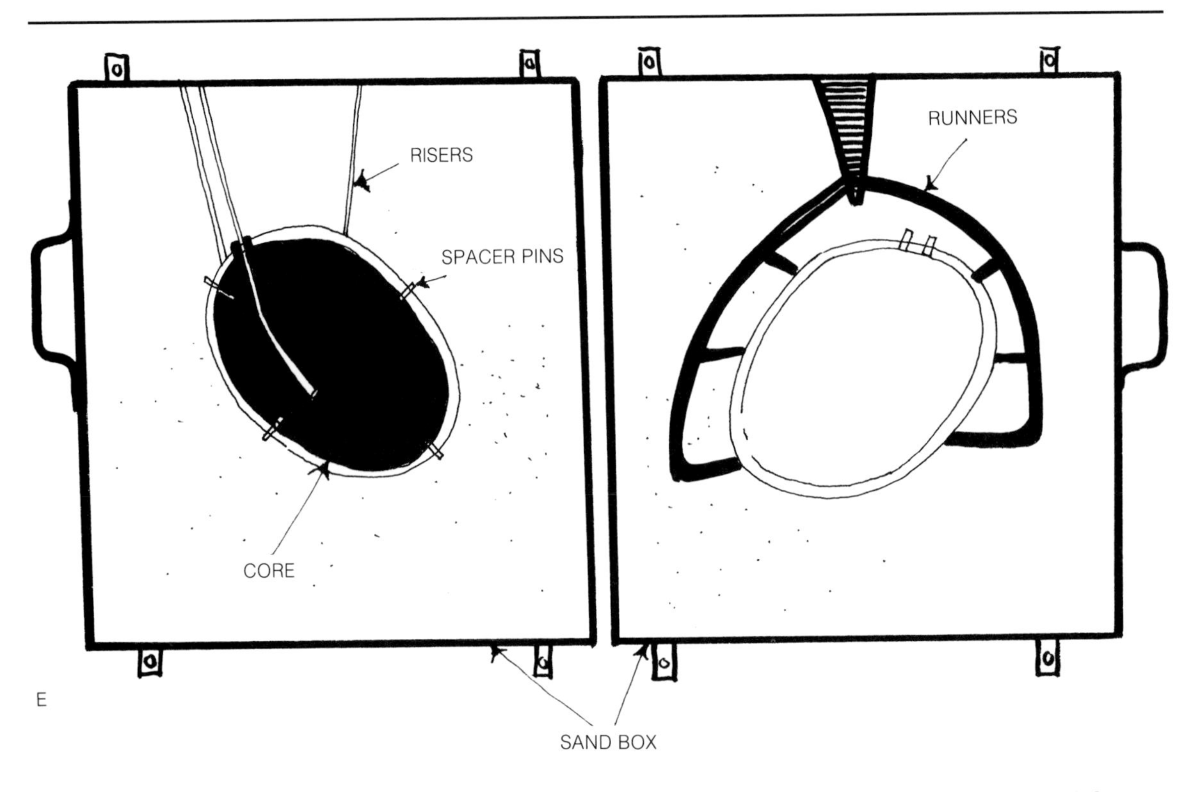

124 Sand moulding, continued from page 137. E and F: a more complex pouring gate cut as channels in the mould halves, risers and runners, and the core held firmly in the mould cavity.

needle-like rods are used to secure the pieces thus keeping them in place during the pour.

5 The *core* for a sand moulding can be made in several ways. The core for a sand casting has to be built up to the same dimensions as the master, the metal thickness being determined by paring down the core. According to the circumstances the core can be made from the same mould as the master if a plaster piece mould is used. It can also be made by modelling a core with sand, constantly comparing it against the master, and trying it in the mould, but this is only possible in the case of simple forms. A core can also be made from the sand mould which has already been taken from the master cast. This is fairly normal foundry practice; a casting is carefully taken from the mould, then equally carefully pared down to determine the metal thickness.

The core is held in place by metal rods (sprigits) which are secured in the core and rest on the rim of the mould. Vents must be made to allow the core gases to escape.

6 The *pouring gate* is cut into the mould whilst the flasks are parted (*124*E). A system of runners and risers is cut into the sand in a series of channels, and a funnel is made allowing proper entry for the metal.

7 When this is complete, the core having been placed, the pouring gate cut, and the extraneous sand dusted out, the mould surface is painted lightly with a graphite. This substance helps the speedy flow of metal into the mould cavity. The metal flasks are then placed together and secured, and the seam edges are sealed, or the boxes packed in sand, to trap any escaping metal. The mould

is now ready for pouring.

Another possible solution to the high cost of bronze casting is a utilisation of industrial sand casting techniques. It is the time of the specialist which costs money, therefore the sculptor himself must become the specialist craftsman. Cheaper castings can be made by designing the master cast, making it possible to use the ordinary industrial foundry with little interruption to the flow of work. This means using the simplest sand moulding techniques. It also means that the master must be made so that it is virtually a collection of patterns to be cast and then assembled to make the final work, the assembly being effected by the sculptor. The pattern pieces should be designed to have little or no undercut and to fit snugly one to another. The foundryman casts from these patterns, casting simply by moulding from the back and front of each piece, the process used being the sand moulding technique (see page 21). The cast pieces are then returned to the sculptor as crude casting, only the pouring gate is cut off and no chasing is carried out. The processes of chasing, finalising a surface, assembling the pieces by welding or having them welded to direction, or bolting them by tap drilling, are all carried out by the sculptor in the studio. Thus the final work is controlled by the sculptor. The master patterns can be retained to make additional castings for an edition or indeed to make variants to the original idea. The bonus to the sculptor in this method is the fact of having produced a sculpture which is durable, for a fair price, and of making a financial return at least equal to his work.

The design and manufacture of the pattern sections require considerable thought, since the quality of the casting will depend on this. Patterns can be made by various methods, e.g. a clay former under a plaster positive, which is then sawn or divided to remove the clay and form the pieces. The former can, for example, be made of expanded polystyrene or sand, and the sections can be cast from a flexible mould taken from a master cast. The traditional technique of making wooden foundry patterns can obviously be used, and in the case of some of the 'cool' contemporary forms, would be ideally suitable.

That the pattern is skilfully designed and manufactured is of prime importance. Secondly, but also a most important factor, is the ability to work on the metal. When the castings are returned to the studio they are crude. Work on them may be difficult and tedious, but I believe that the final image can benefit greatly from this. The *ability* to work on the metal is something which can only be acquired by experience.

Chasing is the term given to cleaning up, polishing and finalizing the metal surface, which is carried out with various hard tempered steel tools. Most sculptors forge and make their own whenever possible, but suitable implements can be bought from a reputable tool supplier. Chisels used for cutting letters in stone, or tungsten-tipped chisels, are suitable for working a metal surface. The cutting edge of such chisels must be well maintained on an oil stone. The work done with these is usually followed by a series of hard sharp files, ranging from a bastard, which has a very coarse tooth, to one with the finest possible tooth. These vary in size and shape according to the work in hand. Surform files in their various shapes and sizes have largely replaced the dreadnought file for use on aluminium. Surform blades are made of pierced hardened steel and have very sharp cutting edges. Waste material passes through the blade so that it does not clog, and these files can deal with soft alloys as easily as they can with wood or plaster. Ordinary files used on aluminium will get clogged up very quickly, as will any drill bits; to prevent this clogging, dip the tools into paraffin (kerosene). It is advisable to make a collection of files and similar tools, as

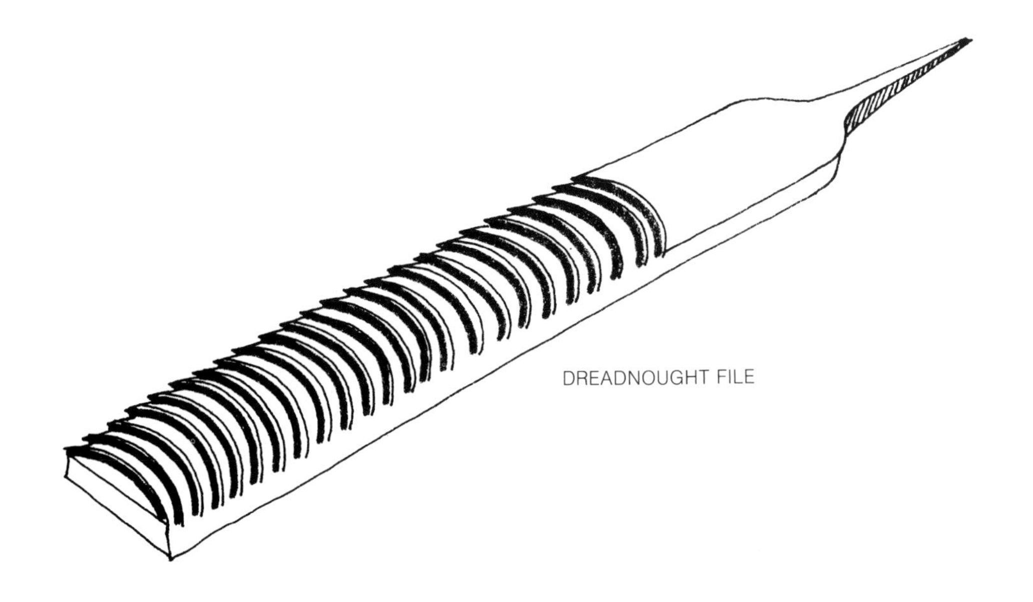

varied in shape as possible, so as to be equipped to deal with any surface from the broadest to the smallest and most inaccessible shape. Files are followed by matting and burnishing tools. Matting tools are simply a variety of punches, with patterned striking surfaces, which vary in size and shape, to make it possible to deal with all surfaces. They are used to matt the metal surface after it has been cut and filed, to make it dense and more uniform. The markings on the striking surfaces assist in making a more uniform surface. The figuring and pattern of these markings are usually devised by the sculptor in accordance with the texture and form he intends to create. After matting, the surface is burnished with hard, polished, metal burnishing tools, which again vary in size and shape. This process is to soften and blend the worked surface with the cast surface. Fine emery papers are also used to give a final softening.

The work on the metal casting includes fixing the various components. Apart from such obvious metal work, all castings will require working to a final form. The sculptures of both Giacono Manzu and Marino Marini retain, in their final form, details that bear evidence to the foundry work; core pin holes are left unfilled, the spots where runners and risers were removed are left unworked. The skill in handling such detail is in retaining these and in no way impairing the visual impact of the form.

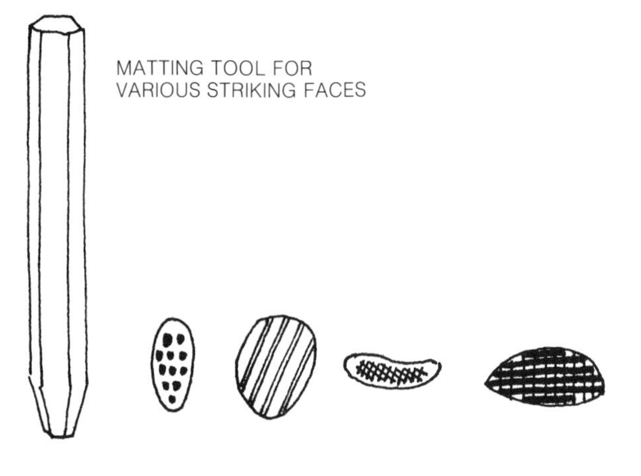

125 A dreadnought file and matting tools.

Cast sections can easily be fixed together by tap drilling and bolting, or riveting, the head of the bolt or rivet being countersunk or worked over. By drilling holes into the matched surfaces, the first a clearance hole, the second to fix the bolt, the sections can be secured. A conventional bolt can be used, or a self-tapping screw, i.e. one that cuts its own thread as it is driven in (*127*).

The sections of castings can also be welded together; this is undoubtedly the strongest and most permanent method of fixing. Welding such castings is amongst the most difficult tasks, however, and the sculptor or student is advised to seek proper tuition to acquire the necessary skill to accomplish such a task efficiently. If this is not practical, then a professional welder should be engaged to do the work. Most foundries employ such a craftsman and, if the work is to be placed in a public site, then professional welding is essential. The area of the weld can be chased and worked in the same way as any other part of the work. Holes in the casting can be filled by drilling the hole out, which can then be tapped and filled by screwing in a threaded metal rod, made from metal of the same type, if not of the same cast, i.e. one of the runners. The screw, once placed tightly, is cut off at the level of the

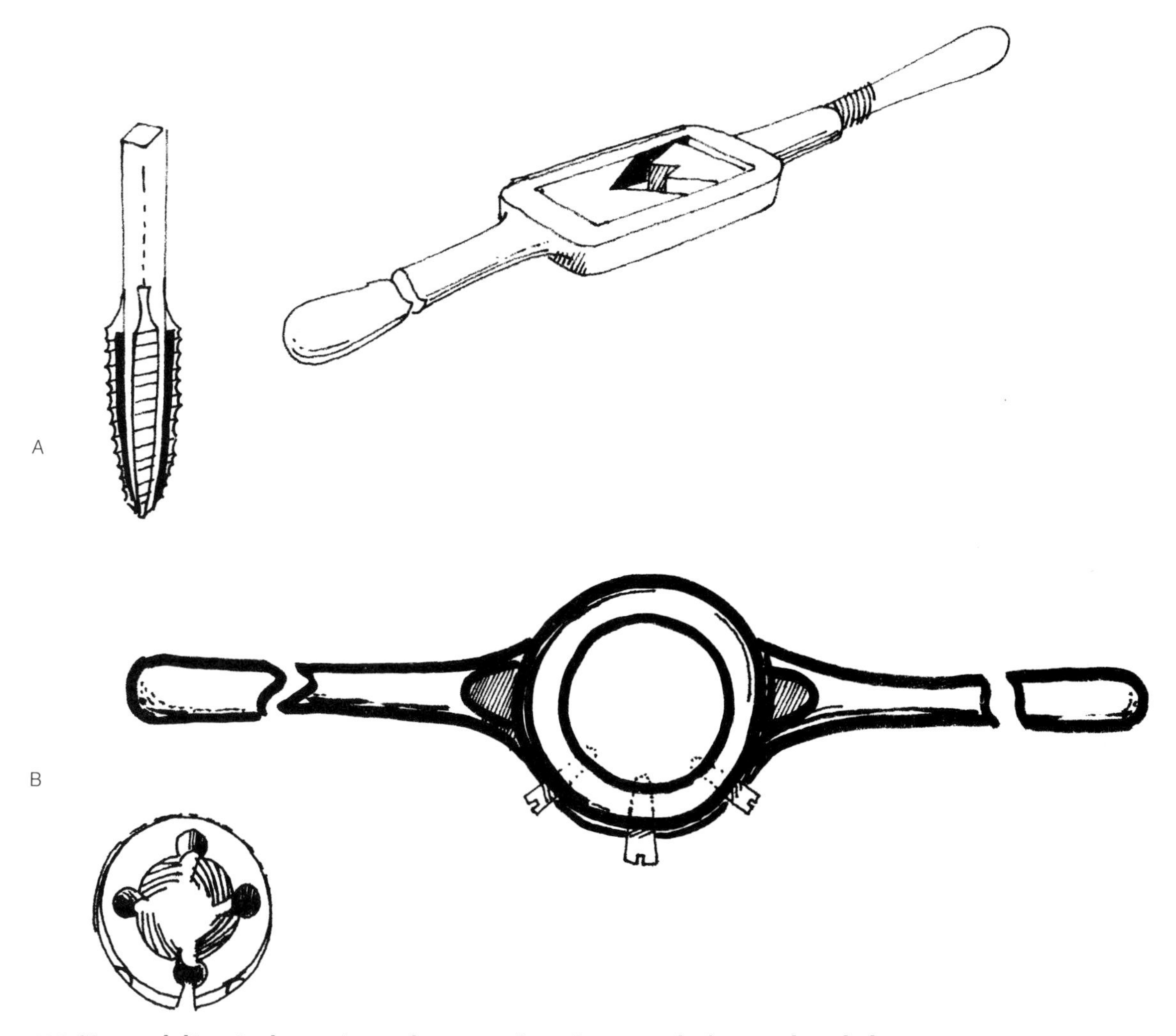

126 Tap and dice. A: the tap is used to cut a thread into a drilled hole. B: the die is used to cut the thread on a shaft to make a bolt.

cast surface; it is then chased, matted and burnished to blend with the rest of the surface. The hole size that can be dealt with in this way is, of course, limited to the drill size. Larger apertures need to be enlarged and shaped to correspond with a previously made metal shape, to make good the fault. This insertion is fixed by brazing, the filler rod used being of the same type and specification as the parent metal. Smaller holes can be filled by brazing also, such brazed areas, great and small, being subsequently chased to blend with the surface.

There are a number of epoxy resin-based metal fillers on the market which are used in industry to repair a broken metal part or to

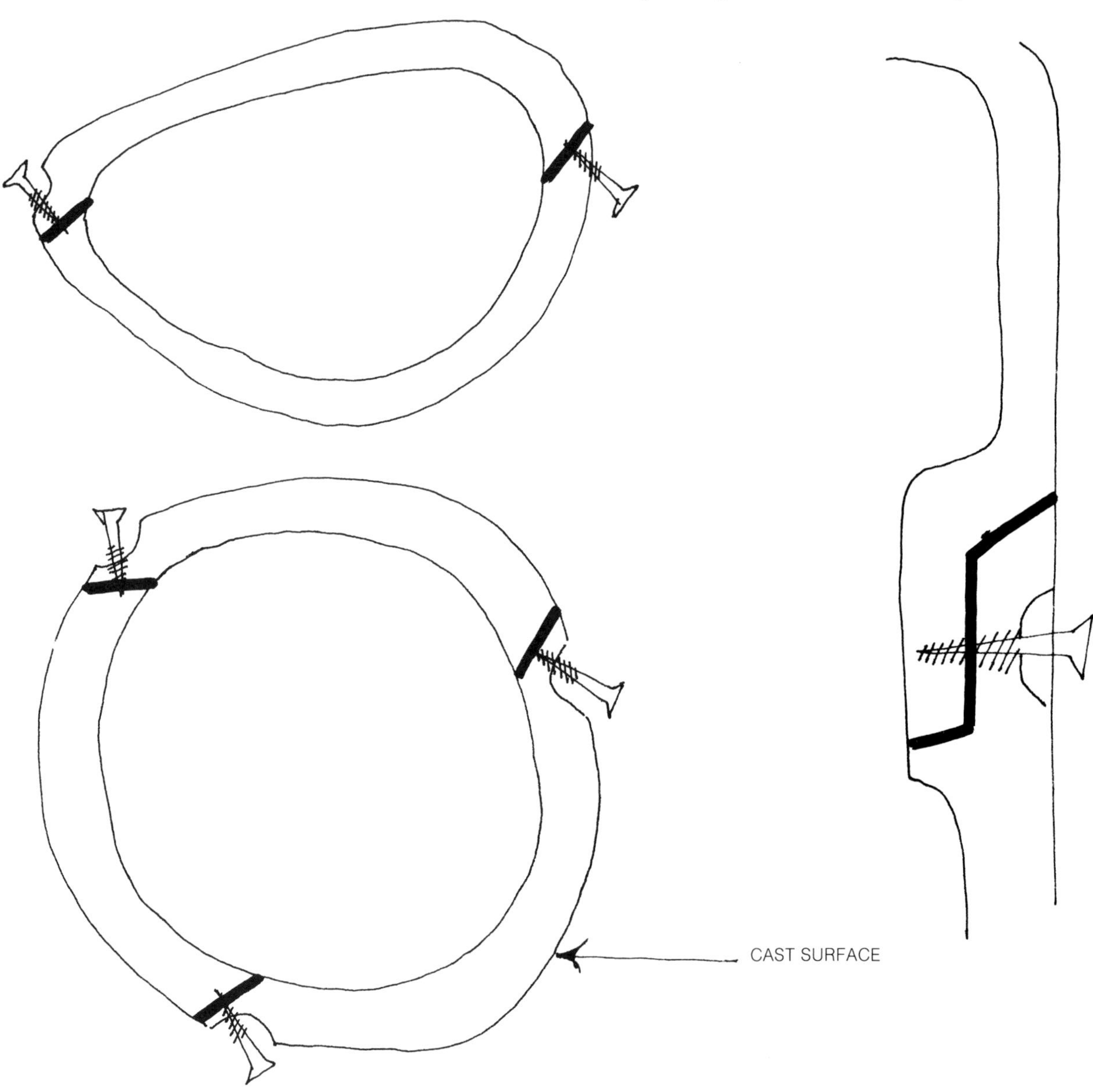

127 Fixing together metal castings using self-tapping screws for sections with matched seam surfaces.

make good a metal casting. These can be obtained in almost any metal character and will prove to be very useful on a casting to fill in a countersunk bolt or screw head, or to fill any gaps between sections when they are bolted together. These I consider equivalent to the various concoctions made up by foundrymen to plug small holes and generally to make good, concoctions that include mixtures such as lead and paint, amber resin, metal filings and paint, solder and paint and even hard wax and metal filings. There is no shame attached to using such synthetic fillers, provided they are used properly as fillers, and not to remodel whole areas of faulty casting. They can be worked like metal, filed, tap drilled, and polished, to blend in with the parent metal. Ideally the surface should be only chased, welded, matted and worked, but it is not always possible to achieve such an ideal.

Furnaces and Melting

One of the most important items to the sculptor who intends to make his own metal castings is the basic equipment. It is all very well to be told 'how', but 'what' is of greater importance. I would emphasise that the foundry business, at whatever level, is not one to be entered into light-heartedly. It involves handling large quantities of heavy molten metal, a dangerous commodity that requires a certain care and confidence. The degree of concentration on what is an arduous, hot and dusty task must be high to achieve results worthy of the time spent. If this is asking more than an individual's temperament permits, it is best left to others.

The basic piece of equipment is the *furnace*, which contains the *crucible*, in which the metal is made molten (*128B*). The furnace is of metal, lined with a refractory material, with space around the crucible, which is placed in the middle. The refractory lining helps maintain heat of sufficient intensity and longevity to melt the required amount of metal, and is directed into the furnace, from its source, through an opening in the side of the furnace wall. The flame enters at an angle to the centre of the furnace, the object being to circulate the flame and heat equally around the furnace, avoiding a direct thermal blast on to the crucible. The furnace has a lid with a system of vents so that the heat within the furnace can be controlled.

It is relatively simple to make this kind of furnace, provided the project is well planned and carefully carried out.

The metal container can be welded and made from mild steel. It can also be made from a large oil drum, or some other cylindrical metal container, preferably of 5 gallons (18 litres) or larger capacity. An opening must be cut into the side of this to allow the flame to enter (*128B*). A refractory lining to the inside of the container must then be made (*128A*). A thickness of 1½in to 2in (38mm to 50mm) should be built up. The refractory material can be of fire bricks, cut and shaped to fit the container, and cemented in position with a fireproof cement (*128D*). Refractory cements can also be used simply to build up the lining, one side of the container at a time, doing the second side when the first has set and hardened. The lining is made all round the sides with an opening, and over the bottom surface, which should include a pedestal for the crucible to stand on. The edges of the lining around the top of the container should be trimmed. The lid can be made separately to fit on to the main part of the furnace, and should also be made in a container and have the necessary air vents. To make a good fit between the lid and the edges of the lining, cast the lid against the top edges of the main furnace lining, using newspaper or tallow as a separator between the cements. An extension ring can be made to fit between the main body of the furnace and the lid; in this way it is possible to accommodate various sizes of crucible (*128B*).

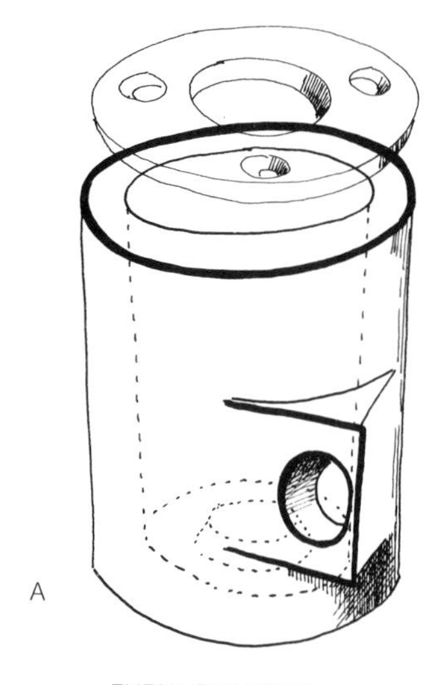

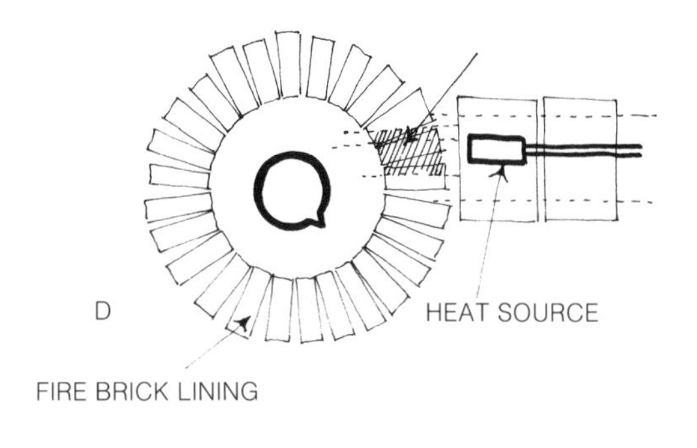

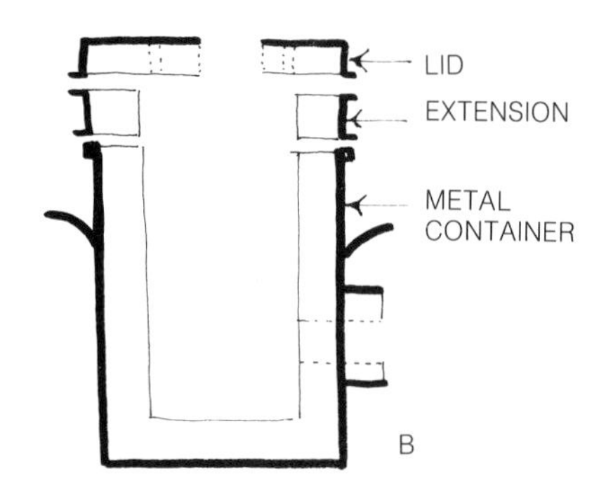

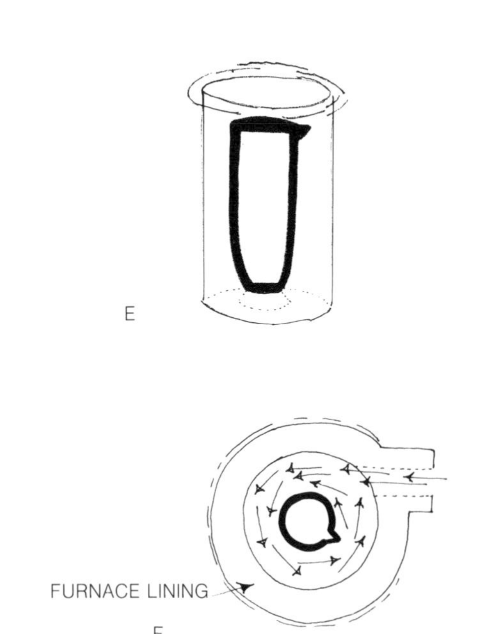

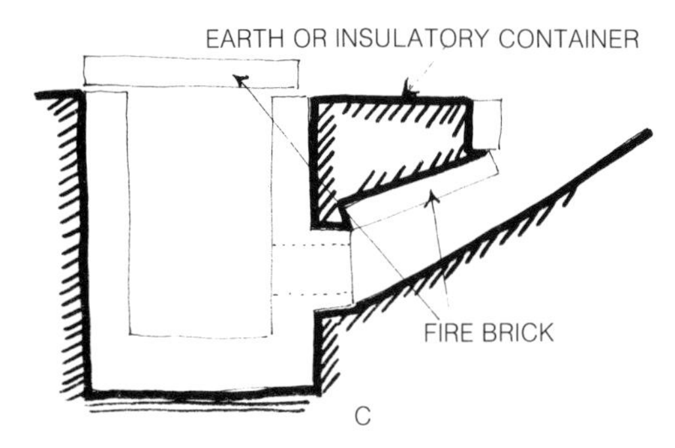

128 Furnaces. A: the furnace lining – this can be cast in a container or built up from fire clay. B: a metal container with an extension and lid. C: the lining buried in soil that becomes the container – note the shaft giving access for the heat source. D: the plan of a fire brick lining in earth to make an efficient furnace. E: the crucible in the furnace chamber. F: the flame movement within the furnace chamber around the crucible.

The simplest way of making a furnace is to dig a pit and channel and line the pit with fire bricks to make the chamber to contain the crucible. The opening through which the flame will heat the furnace is provided in the lining opposite the channel. The fire bricks can be held firmly by packing earth or sand around them, and the lid can be made by simply placing larger refractory panels, kiln shelves for instance, over the top, leaving spaces as air vents. The capacity of the firing chamber can be adjusted to accommodate any size crucible (*128C*).

Another method of making a furnace is to model the refractory lining with a very coarse alumina fire clay, which fires at a temperature higher than the melting point of most metal, i.e. 1300°C (2372°F). This coarse fire clay can withstand the thermal shock of the flame. The lining can be coiled or slab built to provide the necessary features of the furnace, the pedestal for a crucible, openings for the flame, and the lid. The whole thing can then be fired in a kiln to become matured stoneware. With some of the coarser fire clay the lining can be fired in the open air by building a wood fire around and inside it, which is ignited and maintained by stoking. This will reach a sufficiently high temperature to make the body hard and durable and withstand the thermal shock of the furnace flame but will not reach a maturing temperature. This latter stage will be achieved when the furnace is used to heat a crucible.

Stoneware bodies can be used to line a metal

129 As metal is poured from the crucible, slag is skimmed off.

container, and in this case it is advisable to pack an insulator, such as vermiculite or kieselguhr between the lining and the container. Stoneware can also be used in an earth pit, in which case an insulator mixed with sand can be used to make the lining firm and effective.

Crucibles, referred to as 'pots', are specially prepared vessels of fine, high-grade refractory material. These are made to withstand high temperatures, to maintain a maximum strength at such temperatures, and also they contain a specific molten metal poundage. Today these vessels are manufactured for industry and are necessarily of high quality; buying one will prove to be a wise investment, giving long service and the greatest return for outlay. It is possible to make an adequate one, however, from a very coarse alumina fire clay; primitive societies made effective crucibles in this way.

The *source of heat* directed into the chamber can vary. All that is required is that it be of sufficient intensity and longevity to melt the metal in the crucible, the duration of the heat held in the furnace chamber being determined by the capacity of the crucible. Manufactured furnaces are fired using either gas or fuel oil or liquid propane gas (LPG). Specially designed

130 The gas-fired furnace well insulated on sand. The crucible stands in front, illustrating the pouring ring.

131 A gas-fired kiln is used to burn out the wax and prepare the moulds for pouring.

burners incorporating blowers deliver sufficient heat to melt bronze quickly, but such equipment is expensive, and so as with most things sculptors often look for means of providing studio-made products. The best is that manufactured specifically to melt bronze efficiently and safely, and so, bearing this in mind, home made products must be made with care. Costs can be cut, but be sensible about how you achieve such cuts. LPG specific equipment includes a large burner commonly used for tar melting, which is just right for providing the right kind of heat for a bronze casting furnace (this of course will provide enough heat to melt most other alloys). As the gas is delivered under pressure it does not need a blower to intensify the heat. It is wise, however, to make the opening in the furnace side large enough to include space to cause an ingestation of air when the flame is introduced. This can be adjusted to create the correct flame (that which makes complete combustion in the furnace around the crucible). If too great a flame appears above the furnace lid then energy is being wasted where it is most needed: inside the furnace. The flame above the furnace should be green/blue and no higher than 6in (150mm).

Other means of melting bronze I have seen used with varying degrees of success include a 5 gallon (18 litre) blow lamp, running on paraffin (kerosene), which will generate enough heat to melt 20lb (9kg). Two of these lamps, one replacing the other, can be used to melt larger quantities, two lamps being necessary to maintain the heat source without interruption. A paraffin brazing torch also produces sufficient heat, but enough fuel to enable the torch to run at full output for one to two hours is required to melt 50lb (22.7kg).

A great deal of army surplus equipment is suitable for this work and I have seen a furnace heated by an ex-army field kitchen fuel supply.

The most common heat supply used in commercial foundries is gas. It is the most efficient but requires a bellows attachment to produce sufficient heat. Oxy-acetylene burners are also efficient and oil burners can be used, which employ a gravity feed system to supply the heating element. It is advisable to consult the nearest foundry suppliers, army surplus store or engineering supplier, to find out information regarding possible heating appliances, in specific areas.

The *metal* should be cut into small pieces and packed into the crucible, small pieces obviously being easier and quicker to melt than large chunks. It will prove more economic too, because less heat is needed to melt small pieces. Once some of the metal is molten in the pot, more can be added, but do not pack too much in at once. The metal can be bought either from a reliable scrap merchant or from a reputable dealer. The scrap dealer will obviously supply a mixed bag of alloys of a particular metal, whereas the reputable dealer will supply new ingots to a particular specification. This ensures purity of alloy, and consequently control of colour, which is a factor of great importance in formulating patinating recipes – a most difficult process if the alloy is not precise.

When the metal is molten, the crucible is lifted from the furnace, using special tongs and lifting gear (*132*), and slag and impurities which have floated to the surface must be skimmed off with an iron bar before the metal is poured. The pouring is made steadily but quickly; the metal must not be allowed to chill during the pour. When the pouring is complete, i.e. filled to the top of the funnel, the casting must be allowed to cool naturally. Do not be tempted to cool it by pouring on water.

Wear protective clothing in the foundry, especially during the pouring operation. A pair of sturdy boots, a leather apron, leather gloves, and goggles are a necessary precaution. Molten metal, dust, heat and fumes are not to be taken lightly and protection against them should be provided.

When the casting has cooled, break off the mould with a hammer. This will reveal the casting in metal complete with a metal pouring gate. This is a moment of great excitement when the results of the work and effort are seen. Although the majority of casting should prove successful it is as well to anticipate some faulty casts and disappointments. Lorenzo Ghiberti had to make the baptistery doors in Florence, the *'Gates of Paradise'*, twice, because the bronze did not run on the first set. His feelings must have been at the lowest possible ebb when he discovered this, but he made the second doors successfully, and by so doing presented the world with a superlative work. Patience and persistence are virtues that a good sculptor must possess.

Alloys

Alloys are metallic compounds formed by blending two or more elements together, one of which at least must be a metal. Bronze is an alloy of copper and tin, with usually small proportions of zinc, lead or nickel added. Such an alloy has been used by man since the Bronze Age. It is probable that because of the complicated smelting processes necessary to obtain

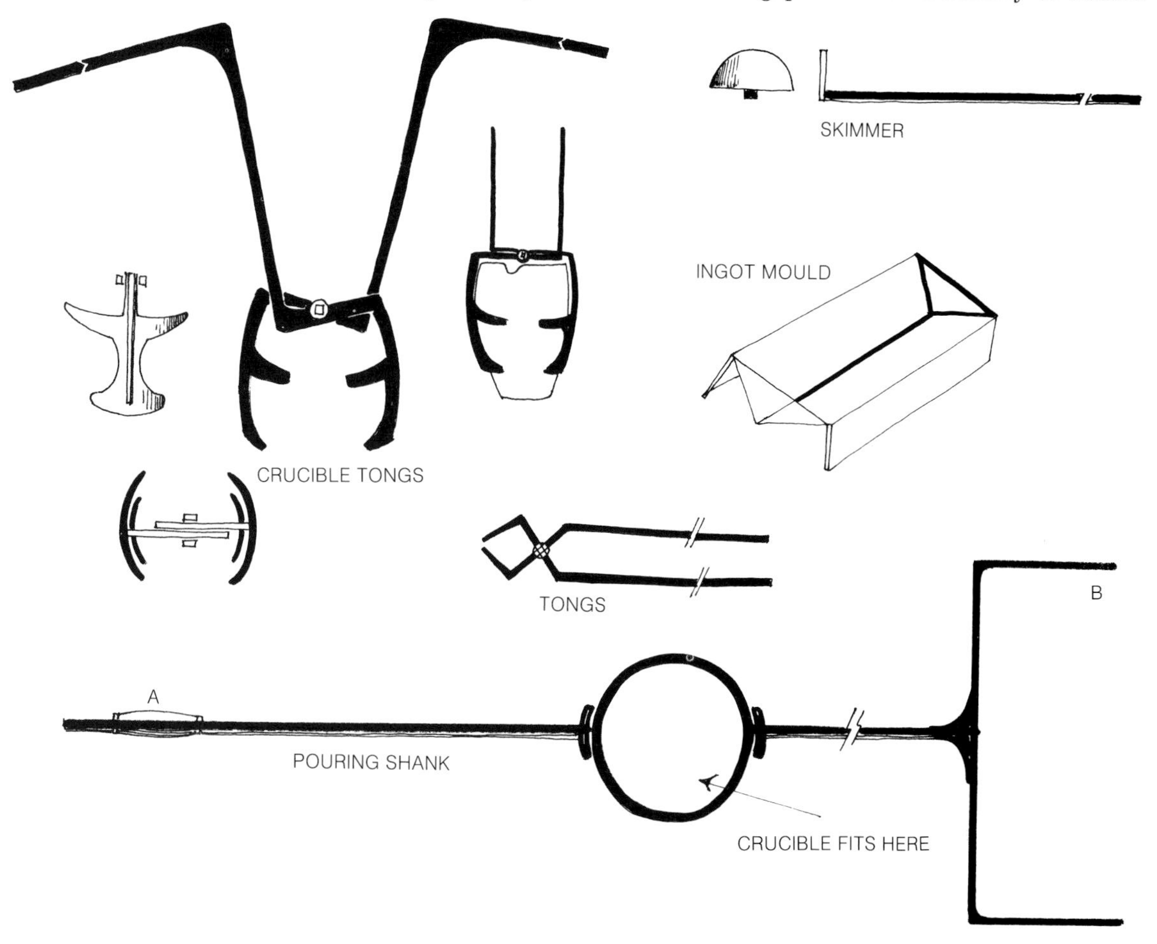

132 Some foundry tools. Crucible tongs are used to remove the crucible from the furnace and to place it in the pouring shank. The pouring shank holds the crucible and in turn is held by two men. One man steadies the pour at A and the other at B tilts the crucible to pour the molten metal. Ingot mould takes any surplus molten metal. Skimmer removes slag.

good working metals from crude ore, metal working was not discovered independently by primitive societies around the world. There are historical implications to indicate that the original discoveries were made in the Ancient East. Full Bronze Ages occur in Egypt, Persia and Iraq c.2000 BC. In China the Bronze Age appears in the second millenium BC. Working in bronze proper, smelting and blending, then spread westwards and northwards from the Near East, becoming established throughout what is now modern Europe by 1500 BC. Since then man has been putting metal to use, and devising alloys to serve particular needs. The composition of some of the common alloys is included in the following list.

Bronze Gun metal originally 90 per cent copper to 10 per cent tin. Now an Admiralty standard metal of 88 per cent copper, 10 per cent tin and 2 per cent zinc. Melting point is 995°C (1823°F). This is most commonly used in sculpture today.

Silicon bronze is an alloy of 97 per cent copper, 2.5 per cent silicon and 0.5 per cent mangenese. It is a good casting bronze because it flows freely. Melting point is 1200°C (2190°F).

Coinage bronze is an alloy retaining a high percentage of copper, but hardened by mixing 95 per cent copper, 4 per cent tin and 1 per cent zinc. Melting point is 1,050°C (1922°F).

Brass is a blend of 70 per cent copper, 30 per cent zinc. Melting point is 1,050°C (1922°F).

Aluminium bronze is an alloy of 90 per cent copper and 10 per cent aluminium. This produces a bronze of a bright golden colour. Melting point is 1,060°C (1922°F).

Aluminium is added in small amounts to certain other metals to improve their properties, or small quantities of other metals and silicons are added to aluminium to produce the aluminium based alloys. Ordinary commercial aluminium is 99 per cent pure, small amounts of silicon and iron being added to increase strength. Five per cent silicon alloys are used architecturally, giving good resistance to corrosion.

Other alloy ingredients include copper, mangenese, nickel, zinc and magnesium. Melting point for all is 660°C (1220°F).

The Lost Pattern Process

Forming hollow castings in metal, using the sand moulding process, has always been of particular concern to sculptors. The problem of making an original form, complete and hollow, to be simply covered with a sand mould, without making complex pieces, has often prevented forms being designed for sand moulding. The possibilities inherent in this coarser material are interesting and, with the advent of foamed polymers, the exploitation of the lost pattern process offers an even wider range of qualities.

The commercial development of lightweight expanded plastic foams, being up to 98 per cent air, has provided the means for evolving the 'lost pattern' process. In many ways this is similar to that of the lost wax process. The original can be made of expanded polystyrene or of rigid polyurethane foam. The metal thickness can be determined during this time, and the form can, of course, be solid or hollow, according to size. The effort and time involved in making a small form hollow may be better spent, however, in controlling the juxtapositioning of volumes, and the overall shrinkage of the metal upon cooling. When encased in a sand mould, molten metal is poured straight on to the original which instantly burns out to be replaced by the metal. In this way many tedious moulding processes are eliminated.

The technique for using these cellular polymers is fairly straightforward. The original or pattern can be made by cutting, shaping, glueing and generally using an amalgam of additive and subtractive techniques. Rigid polyurethane foams can be obtained in various densities, expressed in so many pounds per cubic foot, the greater the poundage the greater the cellular density. Because of this density and their fine cellular structure these foams can be worked to a much finer tolerance than expanded polystyrene. Polyurethane foam can be used as a casting medium as well, but this may also be a fairly dangerous process. The medium is composed of two matched liquids that react when mixed, quickly becoming a stable, rigid foam. Whist reacting, the material produces expansion pressures of up to 15lb per square inch (260g per square millimetre), and high temperatures at the centre of the mass are generated, becoming often as much as 1,400°C (2552°F). So, although the material can be poured as a liquid, later becoming a set solid, it requires a mould capable of withstanding these expansion pressures and heat build-up. The most suitable type of mould for this would be one made of polyester resin, see page 95.

Once the original form has been made from one of these materials, it has then to be prepared for casting.

The original is first fitted with a pouring gate which can be made from strips of the parent material and built up on the master form by glueing. The pouring gate is similar in principle to the lost wax system. A series of runners and risers are also necessary to permit the metal to run and the gases to escape.

The completed original with its pouring gate is then embedded in foundry sand which should be firmly compacted about it. CO_2 sand is commonly used for this purpose because it can easily be hardened by applying CO_2 gas, *in situ*; the mould does not require heating to make it rigid. Greensand (natural bonded) can also be used as it too requires no heating, which is an important factor in view of the nature of the plastic.

Molten metal is then poured directly on to the original. The foamed polymer will volatalise and disappear to be instantly replaced by the metal. When cool the mould can be broken off and the metal casting trimmed and chased to its final state.

This process is yet another in which experiments should be made and developed further.

Lead

This is perhaps the most common studio poured metal since it has a low melting point and can easily be made molten in an iron saucepan over a flame. A gas flame is the most suitable; a gas cooker can be used or bottled gas (cylinder) with an attachment on which to stand the vessel to heat. The process is much simpler than for other metals because lead is regarded as a dead metal that gives off no gases. Consequently it is possible to fill a cavity by simply pouring in the metal, and then allowing it to cool and harden. Some shrinkage occurs when it cools, so allowance should be made for this by placing a large funnel over the pouring entry, which gives a supplementary weight of molten metal. No elaborate pouring gate is necessary.

Although it is possible to make a lead sculpture by simply filling a prepared cavity or mould, students should also be introduced to the principles of casting with more lively metals, by making lead castings using a proper pouring gate system. The resulting lead castings can be cut up and analysed, and in this way invaluable experience can be built up.

The processes used to make lead castings in the studio are as follows:

1 Model the original form from either clay or wax.

2 a From the clay original make a waste mould using 3 parts grog to 1 part plaster of Paris; either the clay wall or the brass shim technique can be used. Reinforce this with mild steel or chicken wire. Remove the clay and clean the mould surface. Fix the mould sections together and seal the seams. The mould must now be thoroughly bone dry before the metal can be poured.

b If the original is wax, this can be covered

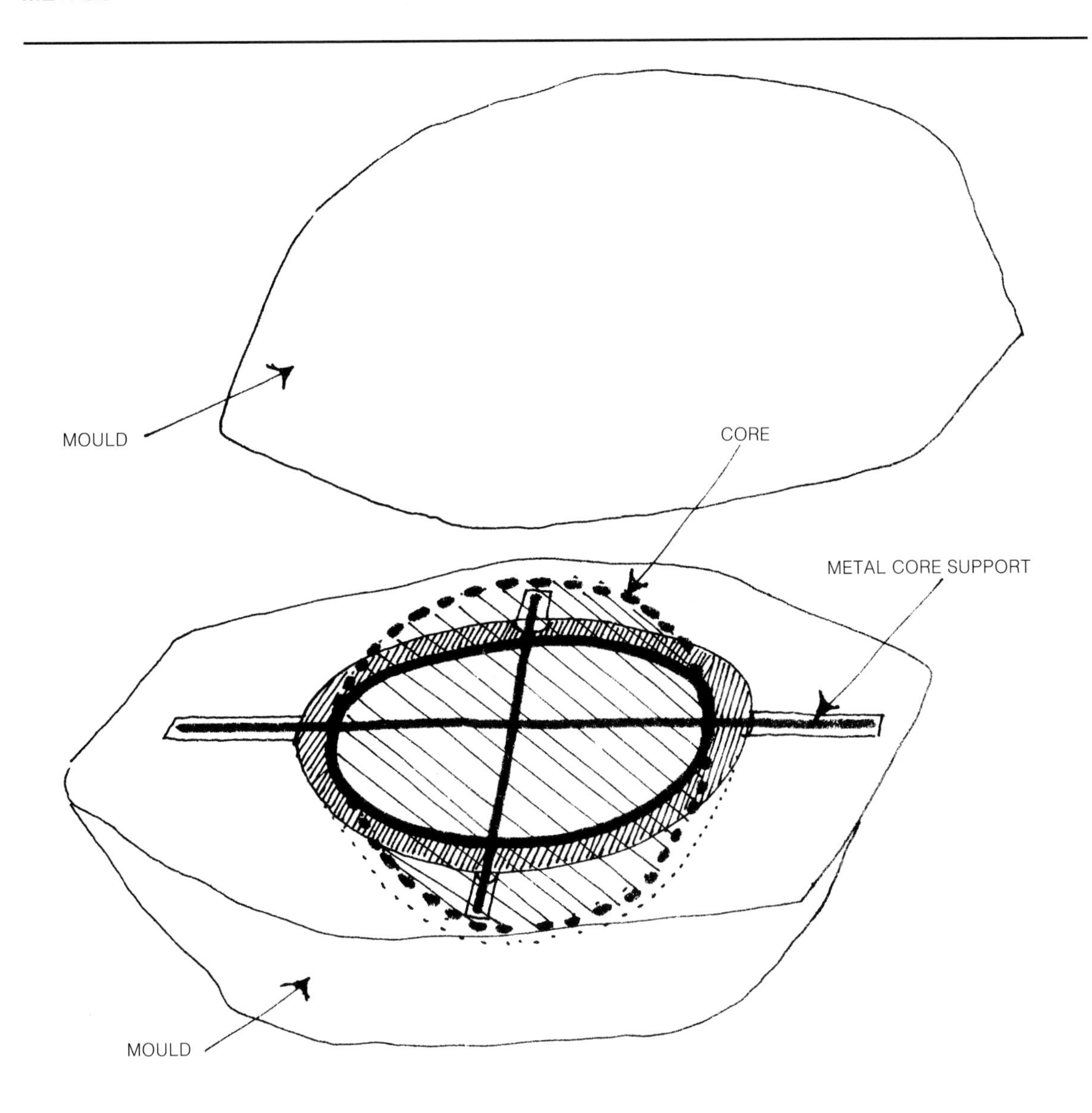

133 A simple kiln made by fixing one container inside another. This is commonly done by securing a five gallon (18l) drum inside a dustbin. An opening is made to permit heat to be introduced to the back of the mould. This kiln is used to prepare moulds in the studio for lead casting.

with a mould, using a 3 parts grog to 1 part plaster mixture. When the mould is placed over the heat source to dry, the wax will be melted out at the same time.

Cores can be placed in both these forms. The wax positive, be it modelled direct or cast from a flexible mould, can be cored simply by pouring, or by modelling over a pre-determined core, see page 131. The clay figure can also be cored simply in the following way. When the waste mould has been made, make in it a clay thickness equal to the planned metal. Cut into the seam edges a number of niches, and design

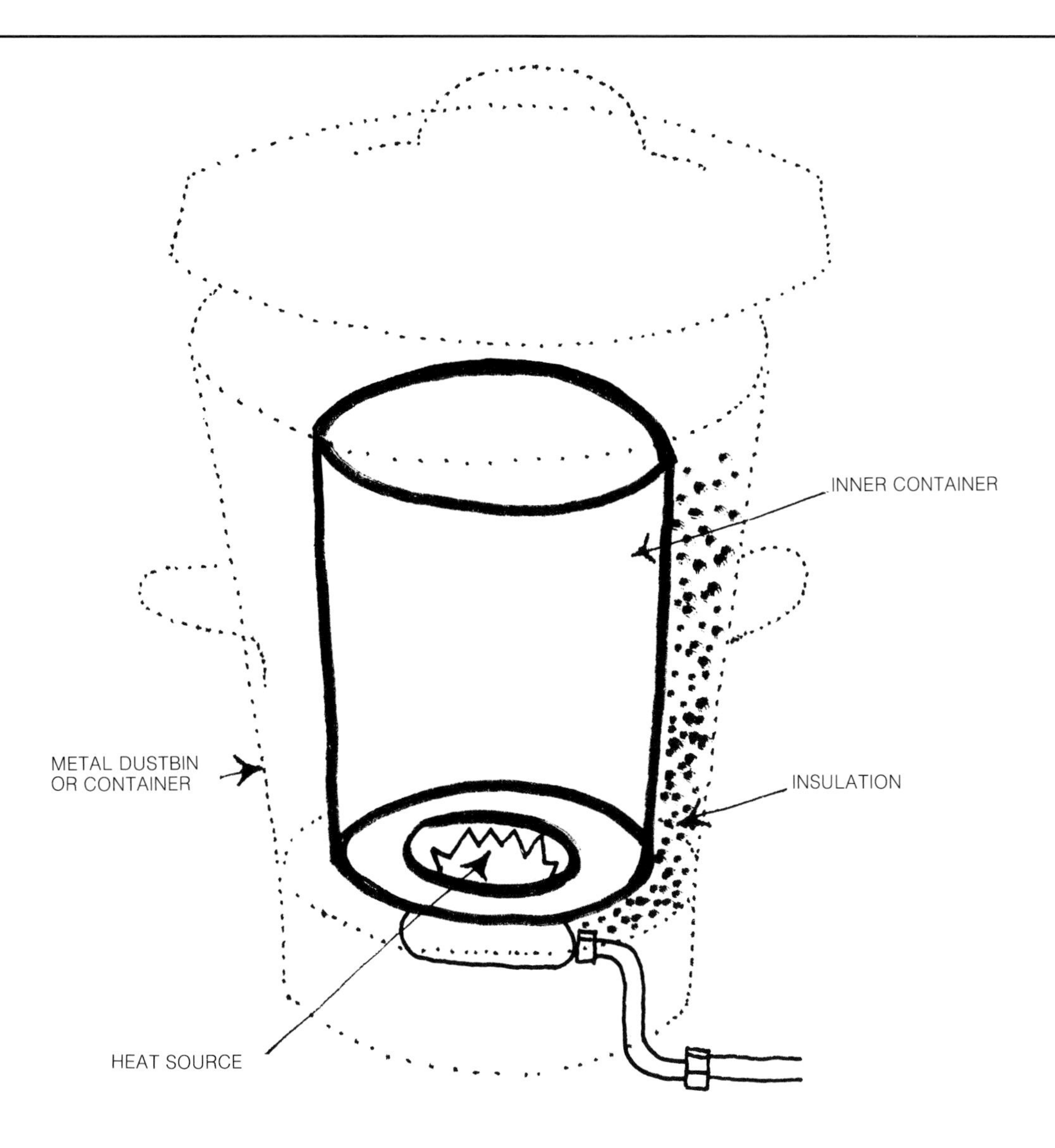

and make an armature to fit into these niches. Place this armature in position, then temporarily fix the mould together and pour in a mixture of 3 parts grog to 1 part plaster, to fill the core space. When this is set open the mould again, remove the core and the clay and clean the mould surface. Replace the core supported at the mould edges by its mild steel armature, secure the mould together and seal it. Then heat it to dry out thoroughly. No moisture should be present in the mould at all. The mould can then be placed inside a makeshift kiln, made of fire bricks over a gas ring, or some such heat source.

3 Another more efficient method is to make a kiln from a metal dustbin (*133*). This is an admirable system, useful in schools for producing small lead castings.

4 When the mould is thoroughly dry, give it almost 12 hours in the kiln. Pack it in a box, pit or bucket according to the size of the mould, with dry sand, taking care not to drop any sand into the mould opening.
5 Melt the metal and pour it to fill the mould. When cool chip off the mould and clean off the cast to finalise the metal surface.

Centrifugal casting is a technique developed in industry to produce very fine hollow castings without making a core. The mould or cavity is made to fit a container, which in turn is mounted into a machine. The machine spins the mould by turning it rapidly at the end of a rotating arm. The molten metal is introduced into the cavity where it is immediately rotated, and in this way, by using centrifugal force, the metal is flung evenly against the mould surface and is retained there until the metal chills. In industry fine quality tubing and cylindrical forms are made in this way. The jeweller also uses this method to produce fine castings of precious metals. With such metals the cast is a very decisive factor, and the less metal that is used the better. The investment material used by jewellers for such castings is a specially fine refractory medium which is poured around the original in its flask or container. No risers or runners are used, but there must be an entry for introducing the molten metal. The mould is prepared for receiving the metal by heating, which removes the wax from the investment at the same time. The machinery necessary for this kind of casting can be expensive, and I advise the interested student to seek specialised tuition in the use of the machinery, and to become familiar with the particular metals and the necessary skills to use them.

134 *Chiavenna Head 1* by Giacometti, bronze. *The Tate Gallery*

VI Machines

There is an enormous number of machines used in industry, and these machines are as complex and varied as the products which they produce. Machines designed to produce items of considerable complexity are in everyday use all over the world, as are simple machines that turn out numerous identical items. It is also this factor that inhibits the use of such machinery by sculptors and designers who are usually only concerned in producing 'one-off' items, or, at the most, a small edition.

There is a justification for mentioning here some of the industrial moulding processes. The items of extreme utility which are manufactured for specific application have about them an austere beauty and unconcern for aesthetics which makes them extremely satisfying to both eye and touch. It is perhaps because the function governs the design and production of working components that inevitable beauty is imposed on them; a factor which bred a kind of inverted self-conscious fashion in the thirties and forties, arising mistakenly some feel, from the German Bauhaus teachings. In the plastics industry, particularly vital today, qualities of form and surface, texture and colour, density and compressive strength, lightweightness and tensile resistance are all put to practical use. It is therefore necessary to indicate the principles involved in some of these. The application by sculptors and designers is, of course, governed by the availability of machinery and raw materials.

The characteristics of thermosetting and thermoplasticity of particular resins should be understood and thought of in conjunction with the machine moulding processes.

My advice is to study the various casting principles involved by means of the following diagrams, and then to arrange visits to plastics manufacturers, trade fairs and exhibitions to see the actual products and machines in operation. It is only by such visits that the full potential of these methods of manufacture and the materials involved can be realised. Processes that offer an extraordinary range of forms, textures and shapes peculiar to the twentieth century, will inevitably be influenced by and will, in turn, influence the design and thought processes of the future.

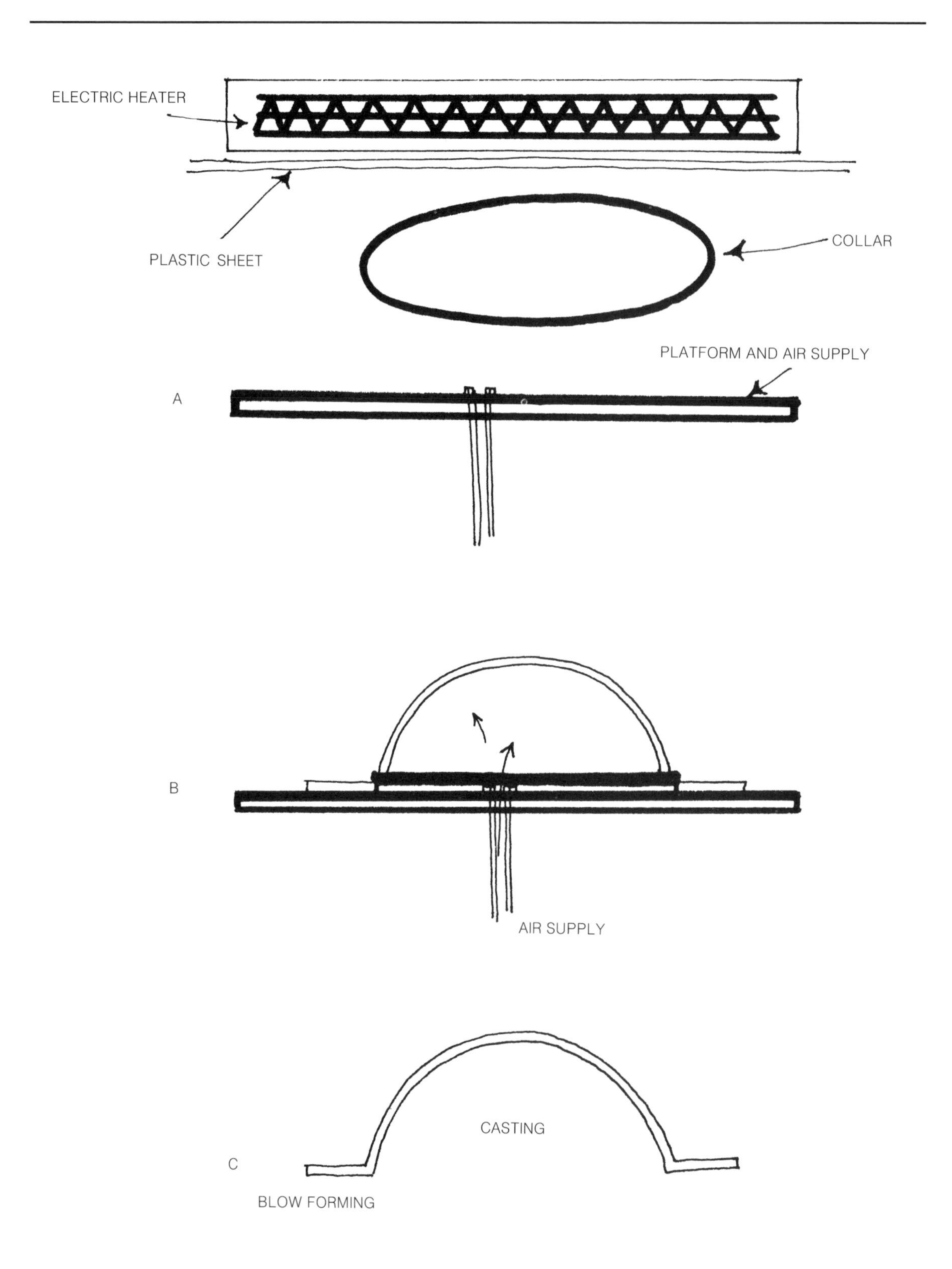
ELECTRIC HEATER
PLASTIC SHEET
COLLAR
PLATFORM AND AIR SUPPLY
A
B
AIR SUPPLY
CASTING
C
BLOW FORMING

135 Blow forming; this is a simple system employed to produce simple blow shapes and is very useful in the studio or workshop. A heated sheet of thermoplastic material is clamped down under a shaped collar, in the centre of which is an opening connected to an air supply. When air is introduced it blows an air bubble, the shape of which can be determined to some extent by the shape of the collar, A and B. The bubble when cool retains the blown shape, C. It is possible to fit a mould or pattern form over the collar to produce a variety of shapes and textures. This provides a useful item of equipment with many possibilities.

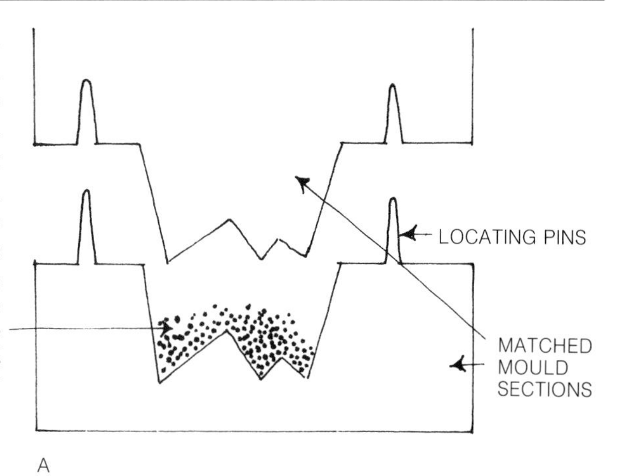

A

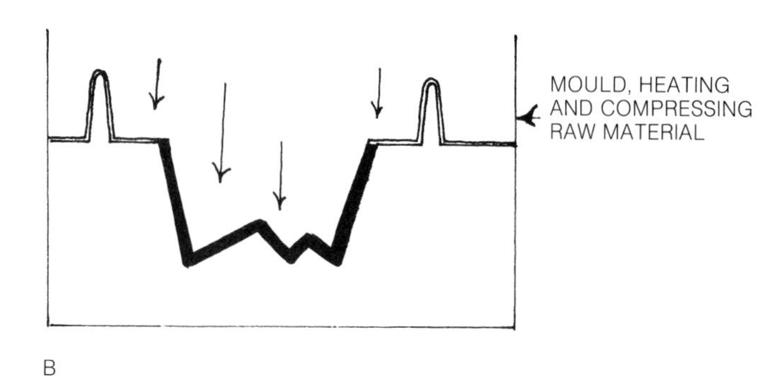

B

136 Compression casting: this is a basic technique using a thermosetting plastic. A: the principle is to introduce into matched mould sections powder thermosetting polymer. B: the mould sections are then brought together, heated and put under great pressure – heating causes the plastic to liquefy and the pressure forces it, whilst in this state, into the mould form. C: the mould is then cooled and parted to reveal a cast form. This method of casting is used to produce much plastic tableware and items of rigid heat resistant plastic. It permits high fidelity casting and a durable substance.

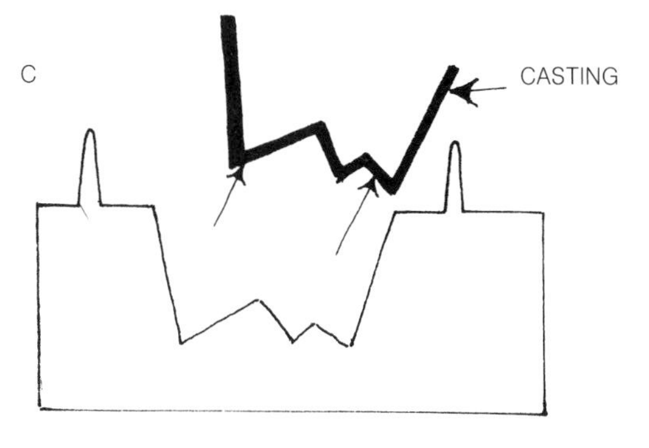

COMPRESSION CASTING

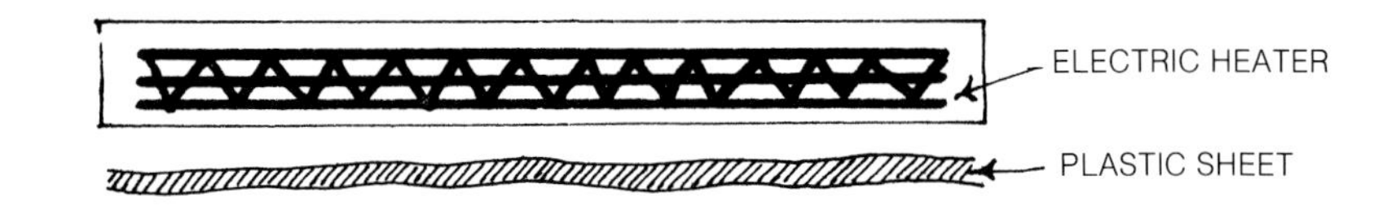

A

MOULD OR HOB

CONNECTED TO VACUUM PUMP

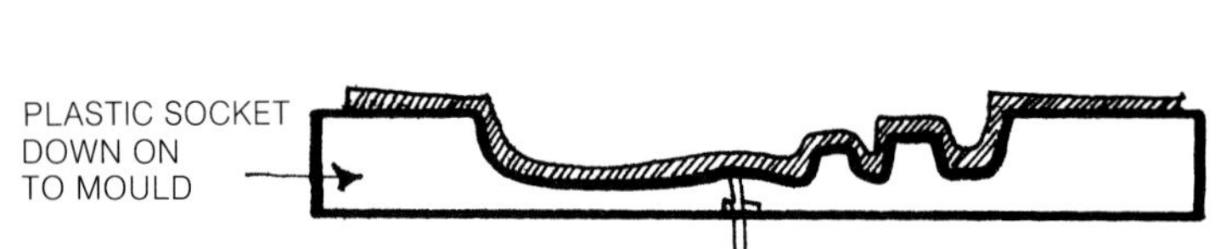

C CASTING

VACUUM FORMING

137 Vacuum forming: this is a method by which a heated thermoplastic sheet is pulled down over a pattern, quickly taking the form of the pattern and retaining this when cooled and hard. A: the plastic is heated over the pattern hob or mould. B: when heated sufficiently this sheet is lowered onto the pattern and quickly pulled down to fill or take the shape of this form by means of a vacuum formed between the plastic and the pattern. C: when cool and hard the casting is removed. This kind of moulding has an endless variety of applications and is used to produce such things as intricate package lining, road signs and items of furniture.

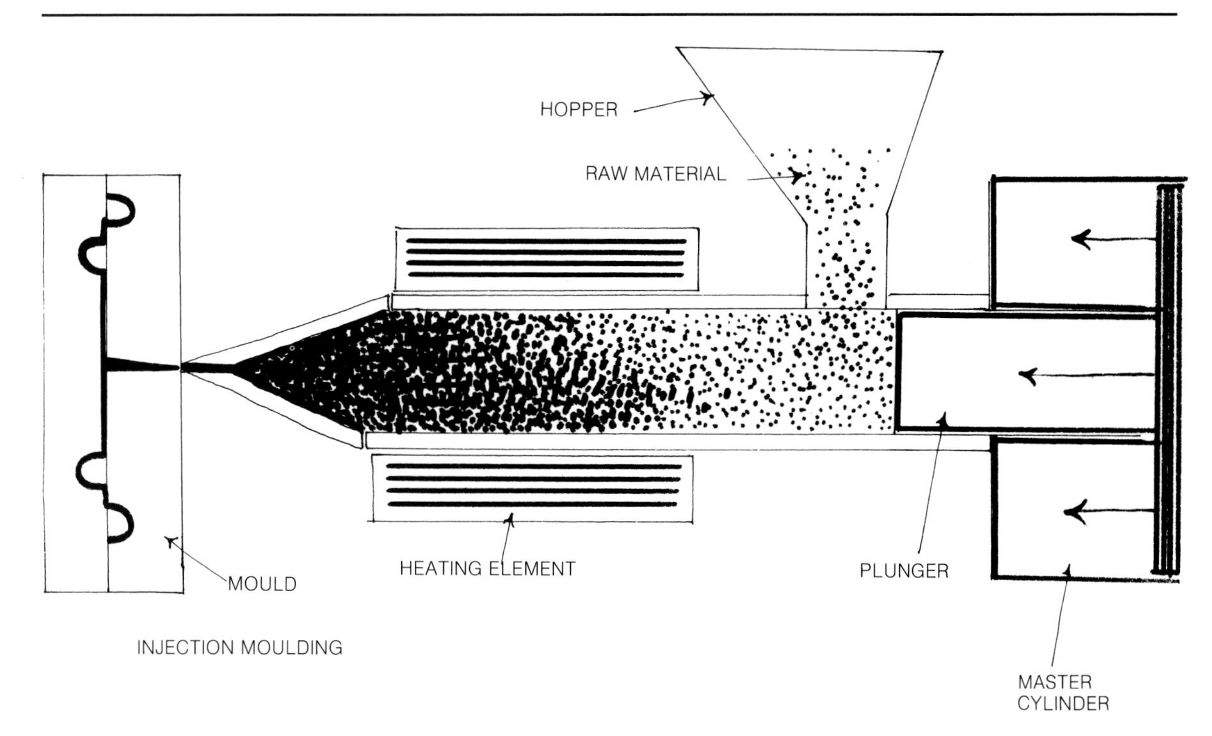

INJECTION MOULDING

138 Injection moulding: the principle of injection moulding is, as the term implies, similar to the action of a hypodermic syringe. The machine is in effect an enormous syringe with a very narrow outlet and a powerful plunger to force, under great pressure, the plastic substance through the outlet. Granules of thermoplastic material are fed into the hopper of the machine and then into the heating chamber. When sufficiently plastic the plunger forces the substance through the narrow outlet that leads the material directly into a mould cavity. Immediately the mould cavity is filled it is cooled by artificial means. This causes the plastic to solidify, becoming hard and insoluble. Evidence of such injection moulding can be easily found in the plastic scale model kit sold in most toy shops. This technique is employed to produce an infinite variety of castings from tiny toys to complex automobile components.

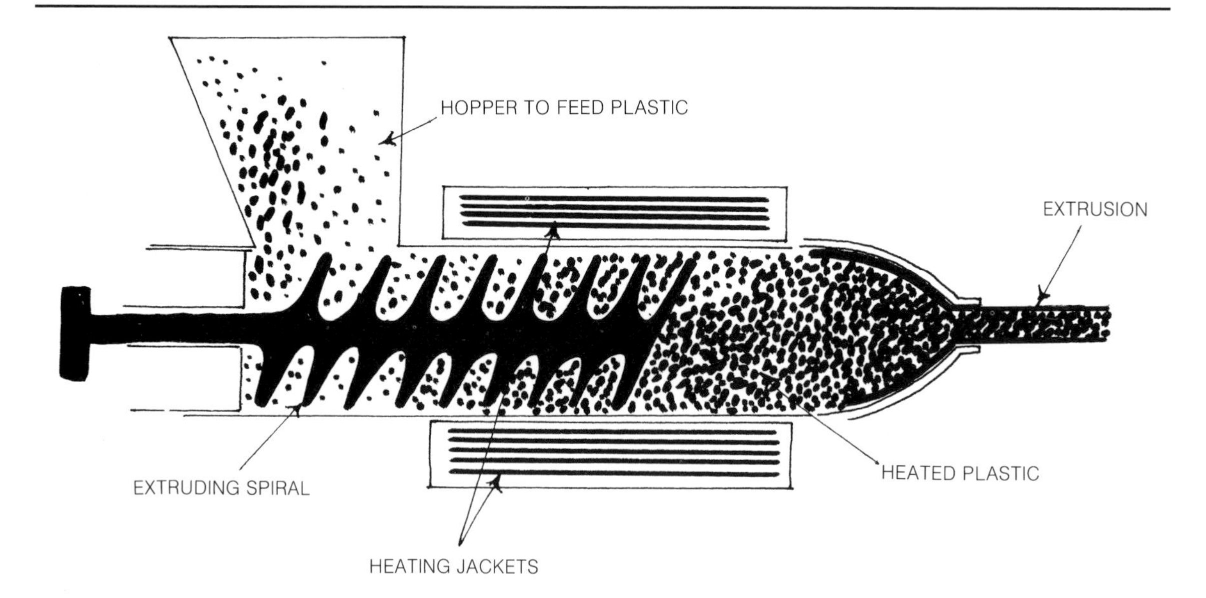

139 Extrusion: this is a method employed widely in industry to produce a continuous uniform form. It is the principle used to make tubing, sheeting and various sections of extruded mouldings. The principle is to feed granules of thermoplastic material into a hopper, which in turn introduces the plastic into a heated chamber where it becomes a thick viscous substance. A revolving spiral forces this homogeneous substance through an opening. The opening forms the pattern for the final form of the extrusion. The plastic is cooled as it is forced through the opening, becoming hard and insoluble and retaining the extruded form. In effect the extrusion plant is like an enormous toothpaste tube, forcing the filling through the narrow opening end.

140 Blow moulding: the principle of this is to force the heated thermoplastic material into a mould cavity by blowing air. A: a method using plastic sheet that is blown into the mould form and heat welded simultaneously. B: the same principle using a heated extruded tube. C and D: the plastic forced to take the mould shape by air pressure. E and F: the blown shape is retained when cooled. This kind of moulding is widely used in industry to produce bottles and various containers. Evidence of such castings is presented endlessly in stores and supermarkets.

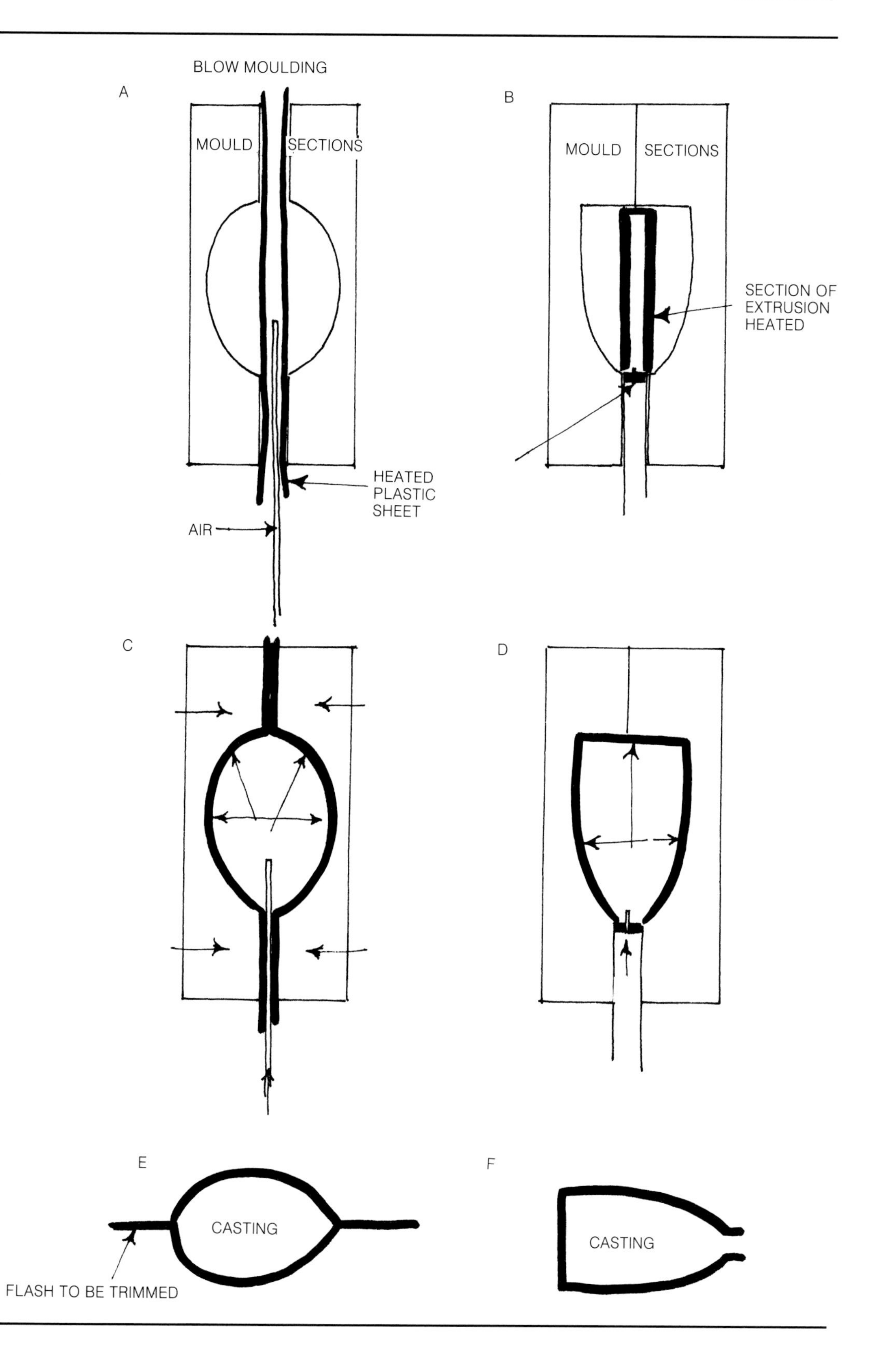
BLOW MOULDING
A
MOULD
SECTIONS
HEATED
PLASTIC
SHEET
AIR
B
MOULD
SECTIONS
SECTION OF
EXTRUSION
HEATED
C
D
E
CASTING
FLASH TO BE TRIMMED
F
CASTING

Glossary

AC Artist copy, usually limited to two castings, made in addition to the declared edition
Accelerator The chemical added to polyester resin to speed hardening; sometimes called promoter
Aggregate The inert granular material mixed with a binder
Alloy A combination of elements, one of which must be a metal
Alumina An earth; the only oxide of aluminium
Aluminium A white ductile malleable metal-

Bastard file Very coarse file for metalworking
Binding agent The material used to create a bond between particles of an aggregate
Brazing Joining two metals with an alloy of brass and zinc
Bronze An alloy of copper and tin

Caps The small sections of a mould removed from the main mould to permit access to the mould interior
Carbon fibre A fibre made from carbon, used in the same way as glass fibre where greater strength is needed
Cast The work that has been produced by moulding; the positive
Casting The process of making an image by moulding from an original
Catalyst The chemical that accelerates a chemical change in other bodies without changing itself; used to effect the cure of polyster resins, for instance
Cavity The mould or space to be filled by the casting substance
Cements The binding agents used in making concrete
Centrifugal Spinning forces flying out from a centre
Centrifuge A machine that incorporates the principle of centrifugal forces
Ceramic Made of fired clay
Ceramic fibre A highly efficient refractory fibre made from alumino silicate. It is used for kiln and furnace lining
Ceramic shell An investment material made from colloidal silica and fused silica aggregate; makes a mould of about ¼in (12mm) thickness with great tensile and thermal strength
Chasing The finishing of a metal surface
Chipping out The act of removing the waste mould
Chopped strand mat Matting made from random glass fibres
Ciment fondu Aluminous cement from alumina
Colloidal silica Used with silica flour to make the slurry investment for ceramic shell; also used as the binder for fused silica to build up the mould
Compressive strength Ability to withstand compression
Concrete A mixture of cement and aggregate
Cradle The wooden support made for a large mould
Crucible The vessel of refractory material in which metals are made molten
Cup The runner cup or pouring funnel located at the top of the pouring gate

Cure Final hardening of a material

Death mask A casting made from a mould taken from the face of a dead body
Draw The taper that enables pieces of mould to draw from the mould case or cast surface
Dreadnought file A file with cutting edges rather than scratching point
Dry mix The mixture of ingredients for concrete, cement and aggregate without water added

Edition The declared number of castings made from the master cast, e.g., $\frac{1}{6}$
Engraving Lines or forms cut into metal surfaces
Exothermic Heat set up internally during chemical changes
Exposed aggregate The aggregate used in a concrete mix revealed by abrading or breaking down the surface fines

Female The mould
Fettling The first working of the metal after casting
Fibreglass Very slender fibres of glass
Filler An inert substance used to add body or colour
Firing The action of heating a kiln
Flash The surplus at a cast seam
Flask The container of metal or some other strong material used to make a poured investment mould known in sand moulding as the sand box or flask
Fluidising bed Aggregate supported by air
French chalk Powdered talc or magnesium silicate
Funerary Items made to be buried with the dead
Furnace The structure in which a crucible is contained to be heated to produce molten metal
Fused silica The coarse aggregate used with colloidal silica to make ceramic shell

Gaiter The system of placing irons to reinforce a hollow casting
Gel coat The first application of resin to a mould that makes the true reproduction of the surface
Gelatine Jelly made from stewed bone matrix; used for making flexible moulds
Glue size A glue or binder made from boiled hide and bones
Grog Ground fired ceramic material added to make a substance refractory
GRP Glass (fibre) reinforced plastic, or glass (fibre) reinforced polyester
Gypsum Hydrated calcium sulphate, from which plaster of Paris is made

Hacksaw A saw used for cutting through metals
Hygroscopic Absorbs moisture from the atmosphere

Irons The term given to reinforcement made from mild steel for moulds and casts
In situ On site
Investment The mould for casting metal sculptures; made from a refractory material, usually grog and plaster

Key The method of making perfect registration; the exact locating of one piece to another, as in moulds
Kiln A furnace or oven for baking

Lamination A thickness built up layer upon layer
Latex A natural air curing rubber; a useful moulding material
Leather hard The condition of the clay before it becomes dry and dusty
Life mask A casting made from a mould taken from the living head
Lost pattern The process of making a metal casting using a foamed polymer positive
Lost wax The process of making a metal casting using a wax positive
LPG Liquid propane gas (bottled gas)

Main mould The part of the mould that contains the bulk of the form, to which caps or

sections of mould are fitted
Male The positive, or cast
Maquette Three-dimensional sketch
Master cast The original work from which a mould is made to make casting; each casting must ideally be a perfect reproduction of the master cast; sometimes referred to as the master pattern
Mortar A mixture of sand and cement
Mould A negative shape or form from which a positive cast can be made; the female

Pack The action of filling a mould by placing the filler
Parting agent Material painted to surfaces that are to be parted (release agent)
Patina The colour and texture of the surface of a sculpture
Pattern The positive form or original from which a mould is made
Piece moulding Moulding by making separate pieces of mould for each undercut form
Pins The iron nails that are placed through the investment mould into the core to retain the space between the two when the wax is melted out
Plaster of Paris Fine white powder that sets hard when mixed with water; made from gypsum
Plastics Synthetic materials, polymers
Polychromatic Coloured sculpture, using many colours
Polymer The substance resulting from polymerization
Portland cement A type of cement, simulating Portland stone
Positive The male cast or pattern
Pour The action of filling a mould with a liquid or molten material
Pouring funnel The funnel placed at the top of the pouring gate to make the place of entry for the molten metal
Pouring gate The system of runners and risers designed to facilitate the easy flow of molten metal through an investment mould
PVA Polyvinyl alcohol, or polyvinyl acetate
PVC Polyvinyl chloride

Release agent Parting agent
Reproduction The quality of cast surface in comparison with the original
Riffler Files with curves working surfaces, used to deal with intricate surface forms
Risers The channels made in an investment or sand mould through which gases and air escape during casting
Roman joints Joints devised and made to produce sections of cast that fit together with perfect register
RTC Room temperature curing
RTV Room temperature vulcanizing
Runners The channels made in an investment mould or sand mould, through which the molten metal runs to fill the mould cavity

Sand moulding The process of producing a metal casting from a mould of bonded sand
Seam lines The evidence on the cast of the separations between the caps, or piece of a mould
Shellac A varnish made from the resinous secretion of an insect, the lac
Shim five-thousandth inch (0.127mm) brass foil used in waste moulding to make the divisions between caps
Shuttering A mould constructed *in situ* usually of wood, to make concrete castings
Silicon bronze An alloy of copper and silicon
Simulate Made to resemble a known quality
Slag Foreign substances separated from the pure metal or alloy when molten; must be removed before pouring
Slip Liquid clay of a creamy consistency. Used for casting
Spatula A simple flat modelling tool of metal, wood or ivory
Squeezing The method of fixing caps to the main mould by laying a thickness of material at the seam surfaces, then squeezing the cap into position

Surform An all-purpose file with a serrated blade to allow the waste material to pass through
Synthetic Artificially produced compounds

Tamping The act of consolidating granular material such as concrete or sand
Tap drilling Making a hole with a thread into which a bolt can be screwed
Technique The mode of execution or skill
Tensile The strength of a material under tension
Terracotta Fired clay
Thermoplastic Becoming pliable when heated
Thermosetting Becoming hard and insoluble when heated
Thixotropic Able to remain on an inclined or vertical surface without draining
Tongs Instruments for holding or grasping, used in foundry work

Vibrating The action of a mould used to encourage air to escape from the casting substance, and to assist the consolidation of the casting material
Viscous A thick sticky consistency

Warning coat The coloured layer of plaster incorporated into the waste mould, which indicates the neariness of the cast when chipped out
Waste moulding Producing a cast by making a mould that is broken away
Welding Joining components of the same metal by causing the edges to become molten, using electrical or oxyacetylene apparatus
Wet mix The state of the mixture of ingredients for concrete, cement and aggregate when water is added

SUPPLIERS

Great Britain

British Gypsum Ltd, Westfield, 360 Singlewell Road, Gravesend GA11 7RY (*plaster of Paris*)

British Oil and Cake Mills Ltd, Stoneferry, Hull, Yorkshire (*rape seed oil*)

Buck and Ryan Ltd, 101 Tottenham Court Road, London W1 (*tools*)

Fiorini Ltd, 7 Peterborough Mews, London SW6 (*bronze casting*)

Fulham Pottery Ltd, 8–10 Ingate Place, London SW8 (*clay*)

J. Galizia and Son Ltd, Chatfield Road, London SW11 (*bronze casting*)

ICI Plastics Division, Bessemer Road, Welwyn Garden City, Herts (*acrylic resins*)

W.A. Mitchell & A. Smith Ltd, Church Road, Mitcham, Surrey (*resins*)

Neogene Paints Ltd, Neogene Walks, Alfred Street, London W2 (*polyvinyl acetate*)

North Road Foundry, Ferrybridge, Yorkshire (*bronze casting*)

Polypenco, PO Box 56, Welwyn Garden City, Herts (*nylon*)

Polyservices Ltd, Taylors Road, Stotfold, Beds (*expanded polystyrene*)

Poth Hile and Co Ltd, 37 High Street, London E15 (*micro crystalline wax*)

Revultex Ltd, Harlow, Essex (*latex*)

Alec Tiranti, 72 Charlotte Street, London W1 (*all materials and tools*)

Vinatex, Devonshire Road, Carshalton Surrey (*PVC moulding materials and equipment*)

USA

Air Reduction Chemical and Carbide Co., 150 East 42nd Street, New York, NY 10017 (*polyvinyl acetate*)

Akron Metallic Gasket Co. (Shim-Pak), 157 N. Union Street, Akron, Ohio (*shims*)

Allied Chemical Co., PO Box 365, Morristown, New Jersey (*glass fibre*)

Carl H. Biggs Co. Inc., 1547 14th Street, Santa Monica, CA 90404 (*polyvinyl acetate*)

American Cyanamid Co., Plastics and Resins Division, S. Cherry Street, Wallingford, Connecticut 06493 (*synthetic casting resins*)

Dow Chemical Co., Midland, Michigan 48640 (*expanded polystyrene*)

Down Corning Co., Midland, Michigan 48641 (*silicone rubbers*)

Cook Paint and Varnish Co., PO Box 389, Kansas City, Missouri 64141 (*synthetic casting resins*)

Fellows Gear Shaper Co., Plastics Machine Division, Springfield, Vermont 05156 (*injection moulding machines and plastics*)

Foster Grant Co. Inc., Polymer Sales Division, 289 N. Main Street, Leominster, MA 01453 (*expanded polystyrene*)

Commercial Paste Co., 925 W. Henderson Road, Columbus, Ohio (*plaster of Paris*)

Furnane Plastics Inc., 4516 Brazil Street, Los Angeles, CA 90039 (*synthetic resins*)

General Electrical Co., Silicone Products Department, Waterford, NY 12188 (*silicone rubbers*)

B.F. Goodrich Industrial Products Co., 500 S. Main Street, Akron, Ohio 44318 (*latex*)

Hysol Molding Powders, 4151 Price Street, Olean, New York (*synthetic casting resins*)
H.V. Hardman Co. Inc., 600 Courtauld Street, Belleville, NJ 07109 (*general elastomers*)
Industrial Production Inc., Davidson, N. Carolina (*shims*)
Jones Dabney Co., 1481 S. 11th Street, Louisville, Kentucky 40208 (*synthetic resins*)
Kindt-Collins Co., 12697 Elmwood Avenue, Cleveland, Ohio (*waxes*)
Lakeside Plastics Co., PO Box 1007, Oshkosh, WI 54902 (*expanded polyester*)
Lester Engineering Co., 2711 Church Avenue, Cleveland, OH 44116 (*injection moulding machines*)
Moslo Machinery Co., 20120 Detroit Road, Cleveland, Ohio 44116 (*injection moulding machines*)
Paisley Produces Division of International Latex and Chemical Co., 1153 Bloomfield Avenue, Clifton, NJ 07012 (*polyvinyl acetate*)
Pittsburgh Plate Glass Co., Figer Glass Division, One Gateway Center, Pittsburgh, Pennsylvania 15222 (*glass fibre*)
Reed Prentice Division of Package Machinery Co., 330 W. Chestnut Street, E. Longmeadow, MA 01028 (*injection moulding machinery*)
Reichhold Chemicals Inc., 525 No. Broadway, White Plains, NY 10602 (*synthetic casting resins*)
Ren Plastics Inc., 5656 E. 30th Street, New York (*synthetic resins*)
Sculpture House, 38 E. 30th Street, New York (*tools*)
Shawinigan Resins Co., 644 Monsanto Avenue, Springfield, MA 01101 (*polyvinyl acetate*)
Sinclair-Koppers Co., Koppers Buildings, Pittsburgh, PA 15219 (*expanded polystyrene*)
Rigron Latex & Chemical Group, 25 Brook Street, Stoughton, MA 02072 (*latex*)
Union Carbide Co., 270 Park Avenue, New York, NY 10017 (*polyvinyl acetate and silicone rubbers*)
The Van Dorn Co., 2685 E. 79th Street Cleveland, OH 44104 (injection moulding machines)
Vermont Marble Co., Proctor, Vermont (*stone and marble*)

141 *Dancing Figure*, James Buth. Bronze height 14½in (36½cm).

Index